A World on Paper

A World on Paper

Studies on the Second Scientific Revolution

Enrico Bellone

English translation by Mirella and Riccardo Giacconi

The MIT Press
Cambridge, Massachusetts,
and London, England

Original edition copyright © 1976 by Arnoldo Mondadori Editore S.p.A., Milan, Italy, published under the title *Il mondo di carta: Ricerche sulla seconda rivoluzione scientifica*

This book was set in Baskerville
by Allied Systems
and printed and bound by Murray Printing Company
in the United States of America

Library of Congress Cataloging in Publication Data

Bellone, Enrico.
  A world on paper.

  Translation of Il mondo di carta.
  Bibliography: p.
  Includes index.
  1. Physics—History.   I. Title.
QC7.B4313   530′.09   80-10674
ISBN 0-262-02147-1

Philosophy itself cannot but benefit from our disputes, for if our thoughts prove true, new gains will have been made; if false, in their disproof the more the original doctrines will be confirmed. Think you rather of the philosophers and be of aid and support to them, because as for science it cannot but advance.

Salviati to Simplicius

Contents

Foreword

My attention was first drawn to this book by its title, for Galileo and
Kepler had both contrasted the "paper world" of natural philosophers
with the "sensible world" of observation at the opening of the scientific
revolution of the seventeenth century. Enrico Bellone, of the University of
Genoa's Institute of Physics, writes of a later scientific revolution, hitherto
less accessible to nonphysicists. The historiographical approach which he
applies, and which in a very real sense he created, is of such intrinsic
interest and wide applicability that I urge its study by everyone concerned
with physics, philosophy, or the history of ideas, scientific or otherwise.

Truly pioneering works of far-reaching implications do not always achieve
wide or speedy recognition, particularly when they go against the grain of
contemporary practice. In my opinion it is by American readers above all
that Professor Bellone's analytical method will be welcomed outside Italy,
and for that reason I take particular pleasure in presenting to them this
English translation of his recent book. Americans share with Italians a
respect for solid results and good-natured skepticism toward grandiose
programs. Given a sound object-lesson, we see how to go on, whereas even
the most profound precepts accompanied by eloquent exhortations may
leave us wondering how to start.

The sciences of heat and electricity grew up in the nineteenth century.
Their evolution has been difficult for nonspecialists to understand because
of the mathematics involved. Accounts of it by specialists versed in
mathematics have tended to neglect all but the strictly logical aspects of its
history. Professor Bellone wants us, if I may paraphrase Galileo, to see
that just as we have eyes to read what nineteenth-century physicists wrote,
so we have brains capable of understanding what they meant.
Mathematics and logic have something to do with questions of meaning,
more or less in the way that locks have something to do with doors. To
open the doors we need keys, and the conventional keys to meanings are
dictionaries. Now, in our natural recourse to standard dictionaries used
by physicists and everybody else, we assume that words are like so many
dried butterflies affixed by pins over their common and their scientific

names, habitats, and so on. But "heat" and "electricity" in the living language of nineteenth-century physicists were more like two butterflies in the park on a summer day, alighting here and there but seldom resting long.

Changing meanings revolt logicians, who see them merely as a source of fallacy in reasoning and to whom the alteration of a definition is a sign of mental instability. Scientists, on the other hand, leave a definition unaltered only as long as it is useful for their purposes. Even mathematical definitions are changed by physicists as they please. But scientists are not thereby irrational; they do not alter definitions whimsically. Neither do they generally overlook the fact that altering one definition implies necessary modification of various others. They just do not bother to write a new dictionary whenever they modify a concept in order to get on with their work.

Historians of science during a period of its rapid change must accordingly become lexicographers, as Professor Bellone maintains. Now, three kinds of questions are confronted by makers of dictionaries: questions of logic, of semantics, and of actual usage (for they omit meanings, however reasonable, that cannot be substantiated by examples, and they include bizarre meanings found in actual usage). Since Aristotle created physics, no physicist has neglected questions of logic, and historians of physics have always been alert to questions of actual usage of terms, especially as a means of identifying precursors. But semantic questions have hardly been noticed in the history of science. It is Professor Bellone who has perceived their importance and in this book pioneers semantic analysis in the history of science. Without semantic analysis the history of science is unlikely to advance beyond its present state.

The word "semantics" does not occur in the book, I believe, and I shall hazard a guess why. It names a field in philology which deals with historical change of meaning, or "semantic shift," originated by Michel Bréal in 1895. But it now also names a kind of fad among logicians in which living words are converted into static symbols, much as Aristotle converted processes of nature into Greek grammar and syntax. History shows this to have been not the best way to further useful science, and what modern logicians have to say about semantics is not likely to aid historians of science except from the time of Aristotle to that of Galileo. It certainly has no use for students of heat and electricity in the nineteenth century.

On the other hand, the methods of comparative philologists in dealing with semantic shifts of ordinary words over periods of time and among related linguistic communities are exactly what should be applied to the history of science during its periods of rapid change. Such periods are appropriately called "revolutions" when the changes are so great and so rapid as to occasion discomfort and resistance. I agree with Professor Bellone that the seventeenth century and the nineteenth saw scientific revolutions properly so called, though I should name these the third and fourth, not the first and second, which I see as having occurred in the Alexandrian period and during the fourteenth century.

Electromagnetic theory and thermodynamics in the nineteenth century were not unfamiliar to me before I read this book, but my understanding of their origins and developments has been immeasurably increased by it. I believe that this book can throw valuable light on them for anyone, and along with that, still more valuable light on how to advance the historiography of ideas in whatever area. What it will not do is to provide any formula by which global questions can be answered, or any royal road to confident knowledge without arduous and painstaking work. For if I read Professor Bellone aright, he would agree with me in paraphrasing Galileo at the end of the first day of his *Dialogue:*

There is no event in the history of science, not even the least that exists, such that even the most ingenious theorist can ever arrive at a complete understanding of it. . . .

But what sublimity of mind was his who dreamed of finding means to communicate his deepest thoughts to any other person, though distant by vast intervals of place and time . . . by the different arrangements of twenty-six letters!

Stillman Drake
Toronto, September 1979

A World on Paper

I discorsi nostri hanno a essere intorno al mondo sensibile,
e non sopra un mondo di carta.

Galileo, *Dialogo sopra i due massimi sistemi del mondo*

Preliminaries

There is a temptation hidden in the pages of the history of science—the temptation to derive the birth and death of theories, the formation and growth of concepts, from a scheme (either logical or philosophical) always valid and everywhere applicable. No doubt the history of science would be easily told if its course were indeed governed by such simplicity. If this were the case, however, the description of the complex process by which man restructures his knowledge of the world according to the answers Nature yields to scientific inquiry, as well as to the problems theories pose about themselves and their predecessors, would be reduced to a list of repetitive examples. Instead of dealing with real problems, history would then become a learned review of edifying tales for the benefit of one philosophical school or another. It would also have very little to offer to the scientist who is currently engaged in research or to the layman who lives his everyday life surrounded by the products of that research.

On the other hand, it is not only difficult but disturbing to resist the temptation of the scheme. Historians worry, with some justification, lest by giving it up they would lose along the way the sense of history itself and shatter the rationality of the process by which man has endeavored over the centuries to create a scientific view of the world. They fear that by questioning the scheme they will be establishing a dangerous and disorienting premise, leading ultimately to the entanglement of all the threads of historical discourse—or, in other words, contributing to a devaluation of the objectivity of knowledge, which is the favorite target of every kind of irrationalism.

Those of us who nevertheless wish to re-examine the history of science (and to look into the nooks and crannies of our not-so-ancient profession) should bear in mind that in every field of inquiry the failure of a method or the weakening of an explicative formula is not the result of an incurable flaw of reason itself. On the contrary, it is a positive sign, pointing to the inexhaustible wealth of the objects and the theories on which reason, amidst all the contradictions, learns to operate ever more effectively. Thus the need for a re-examination of the history of science stems not from a decadence of the profession of historian but from the new and fertile questions that this profession has been able to formulate about itself and its field of research.

Having said this much, it becomes simpler to give a general idea of the motivation behind this book and of the structure chosen for it.

The motivation may be explained by saying that the usual historical analyses of eighteenth- and nineteenth-century physics and mathematical physics are vitiated by a philosophical and historiographical abuse of the term *mechanism:* to this term is widely attributed a theoretical charge of such magnitude as to cause real and substantial deformations in both the history of physics and in physics itself.

The term mechanism generally denotes a set of philosophical beliefs and methodological rules embodied in a nearly unalterable nucleus that is supposed to define, as well as guide, all theoretical and experimental research in the natural sciences from the beginning of Galilean and Cartesian thought to the end of the nineteenth century. Concomitant with the use of this term is the opinion that the roots of contemporary physics (in its relativistic and quantistic components) lie in a philosophical verdict against mechanism itself and in a deep-seated crisis of physics, which verdict and which crisis occurred in a relatively short period of time around the turn of the century.

When we re-examine the writings of eighteenth- and nineteenth-century physicists, however, we find that they do not fit this label of mechanism. We soon realize that these writings are not merely extensions of the mechanistic programs of the 1600s and the early 1700s but also embody an inner criticism of those programs, speculation on their limitations, and a growing awareness of the need for a radical change in the mechanistic view of the world. It is in fact in the eighteenth century that the rational investigation of natural phenomena puts into question the belief in a universe understood as a cosmic clock without history, thus establishing the premises for the overturn of the scientific view of the world. With increasing assurance, there emerges from this revolution a new view of the world, wherein events no longer repeat in accordance with cyclic models and are no longer governed by unalterable rules—a world subject to an evolutionary process that affects both organic and inorganic forms of matter.

This new vision revealed serious problems and contradictions within the mechanistic legacy. And it is fair to say that it was the intense effort to elucidate these problems and contradictions, together with the reflections on scientific explanation inspired by such efforts, that laid the foundations for the second scientific revolution.

This revolution, whose origins we place between the end of the eighteenth century and the first decades of the nineteenth, started with the new theories of thermodynamics, radiation, the electromagnetic field, and statistical mechanics. All these theories raised questions about the structure of matter and the very meaning of physical law, and thus influenced, in different form and measure, both the Galilean–Newtonian tradition and the other sciences. We need only recall the complete rethinking on the foundations of mechanics that took place in the nineteenth century and the effect that the new physical view of the world had on other disciplines, such as biology, chemistry, and geology.

It should be noted that the so-called mechanistic conception, weakened in its philosophical scope and eroded within nineteenth-century physics, continued to exert some influence on certain trends of biological and biophysical research. But the survival of mechanism in different forms and stages should be seen as a historically understandable indication of the upheaval occurring in the rational understanding of the world, not as a proof of the dominant role played by an unalterable philosophical nucleus in all the classical sciences of the nineteenth century.

The presence of such varied stages of mechanism and their manifold developments support the view that revolutions are extremely rich and complex processes rather than abrupt jumps and turns. This point needs to be emphasized, for the long labors of classical physics cannot be reduced to a sudden failure occurring some time between the end of the nineteenth century and the first few years of the twentieth.

The latter is unfortunately a widely held misconception, and its victims are apt to envisage a history of physics whose fabric is rent at the turn of the century and to see the new theories of relativity and quanta as sudden eruptions issuing abruptly from a real fissure in history.

This historiographical delusion has its roots in a specific juncture in philosophy. Unable to keep pace with the radical changes wrought by classical physics in the Galilean–Newtonian legacy, many philosophical schools attempted to overcome their tardiness by ascribing their own crisis to physics. Having loaded the term mechanism with enough latitude to cover the entire course of physics from Galileo to Planck, most twentieth-century philosophy interpreted the new sciences as a sudden turn from, and a definite break with, "that" mechanism, although that mechanism existed only in the philosopher's mind. A majority of philosophers could then decree that the twentieth century began in a climate of unprecedented crisis in scientific thought owing to the failure of classical physics. In this fashion, philosophy managed to turn a rational revolution into a downfall of knowledge, a collapse of objectivity, a crisis *tout court*, while portraying itself as the discipline that makes physicists (and scientists in general) aware of what they are doing and of what they must not do.

In reality, there occurred a scientific revolution that produced a philosophical crisis. But philosophy reversed this relationship so that it could be described as a philosophical revolution that provoked a scientific crisis. Historians of science were asked to follow like water bearers and to manufacture examples that would support this reversal.

Our intent to vindicate the revolutionary character of nineteenth-century classical physics does not necessarily entail a devaluation of the innovative character normally ascribed to the theory of relativity and to quantum mechanics. Rather it leads us to consider the physical sciences of our century as the most problematic products of the revolution that began between the end of the eighteenth century and the first decades of the nineteenth. In other words, it leads us to conclude that the second scientific revolution is still going on today.

Starting from this point of view, there are essentially two sets of problems that need to be investigated. One set of problems concerns the possible correlations between the second scientific revolution and the changes that occurred in man's knowledge of the world during the long period of time from Copernicus to Newton. The second set includes the problems we encounter in refuting the opinion that present-day physics is the strange daughter of a philosophical verdict against mechanism and the foundations of classical physics.

All these problems, by the way, may concur to inspire new reflections about a different question, namely, to what extent contemporary sciences are Galilean sciences. Such a question is clearly beyond the scope of this book. However, since the intent of these preliminaries is to indicate the motivation for the pages to come, it did not seem inappropriate to state briefly some of the reasons why we were induced to reflect on this point.

Lastly, with regard to the structure we have chosen for this volume, we will say that in the first place it is meant to conform to our intent to re-examine the history of some particularly significant phases of classical physics from 1750 to 1900, paying special attention to the relationships between mathematics and experience. We have also attempted to clarify the process by which the relationships between mathematics and experience have been historically determined by the questions that physical theories raised about themselves and their predecessors, and to show that such a process cannot be reduced to an unquestioning acceptance of methodological precepts. The above implies that the "paper world" of theory experienced a very lively growth between 1750 and 1900.

This point of view has certain implications for the way in which the history of physics (and of science) should be done, and provides an opportunity to take a stand on certain assumptions of objective knowledge. These points are dealt with in the first and very brief part of the book, where we introduce the idea of a "dictionary" and suggest that it might help to solve some topical problems in the history of physics without appealing to the traditional distinctions between historical continuity and discontinuity, internal and external history, but rather by aiming the historian's tools at a reconstruction of the weak and strong interactions that occur and develop in our understanding of the world.

The concept of dictionary is then applied in the second part of the book to suggest specific solutions for certain local problems. The five essays in the second part were not written independently, even though their order violates the expectation of a chronology by successive steps. The bond that ties the essays together is not the chronology of the arguments discussed in them but the genesis of the arguments themselves and their critique. Mathematical physics does not grow by a cumulative process but by a process of restructuring. The path of reason through the world of things and theories does not follow a linear trajectory, and the reason for this is that neither the unexpected answers Nature gives to her students nor the questions raised by each successive theory obey rigid formulas; on the contrary, they appear to run counter to the expectations of all those who would like the world to be constructed in a banal fashion and explanations always to be definitive.

The fifth essay should be read as a conclusion; it suggests that we rewrite the history of classical and contemporary physics without dividing it into two distinct branches separated by the imaginary fissure that, according to certain philosophical arguments, should be placed at the turn of the century. If this suggestion finds confirmation in further studies, it will not be inappropriate to say that the history of physics is rich in significant problems and stimuli that are relevant not only to scientists and philosophers but to all who endeavor to achieve a wider diffusion of scientific culture among the masses.

Several people have had a significant influence on my thinking during the writing of this book through articles, conversations, and letters; others have given me precious advice after reading the manuscript. I wish to express my gratitude to Antonio Borsellino, Ettore Casari, Maria Luisa Dalla Chiara Scabia, Ludovico Geymonat, Edgardo Macorini, Felice Mondella, Carlo Montaleone, Paolo Rossi Monti, Silvano Tagliagambe, Sebastiano Timpanara, and Giuliano Toraldo di Francia.

I
REFLECTIONS ON THE HISTORY OF THE PHYSICAL SCIENCES

Gurdulú swallowed a pail of salt water
before he realized that it was not the sea that must fit into him
but he into the sea.

Italo Calvino, *Il Cavaliere Inesistente*

1   *The Scientist's Dictionary*

Scientific Language and Translation

We owe to the wit of Mark Twain the description of a certain manner of reconstructing history: "In the real world the right thing never happens at the right time and the right place. It is the task of the historian to remedy this error."[1]

No doubt, Twain's comment is very comforting to those who judge history to be a pile of debris. If the real course of events were a senseless string of whims and tricks, it *would* be the duty of the historian—and of the philosopher—to intervene in order to establish order out of chaos, remedy errors, and point a finger at the pitfalls that have deceived men and devastated theories in the past. Confronted with a reality disarranged by unforeseeable wiles, the historian in general, and the historian of science in particular, would first of all decide which reconstruction is the most welcome or desirable, and then eliminate the inconvenient facts—the errors. The results of this twofold procedure would make an edifying narrative, particularly for those philosophers who, having already decided how history should be, only ask from historians some comforting examples.

The power of examples is undeniable. Let us assume, for instance, that a historian of physics wishes to investigate the state of knowledge concerning the motion of fluids in the second half of the eighteenth century. He would find that in 1777 French researchers published an important monograph on the subject. In this monograph certain experimental results are described and conjectures enunciated that undoubtedly concern the mechanics of fluids. In the same monograph, however, there appear statements that have nothing to do with fluid mechanics. For instance, one reads that "there is no science without reasoning, that is, without theory." This statement may be judged valid or not, but in any case it has no immediate connection with the laws describing the motion of water in canals. Moreover, some historical facts come to light that bear no relation to either the mechanics of fluids or the importance of theory but are nonetheless connected with the 1777 monograph: this study turns out to be the product of work commissioned by government agencies and therefore implies the existence of particular needs of a social and economic nature.

In short, the monograph in question answers needs of a practical nature, uses arguments of a philosophical nature, and reflects a certain level of scientific knowledge. Well, then, which of these three aspects is the most relevant?

There are several answers to this question. One answer would be that the historian of science should focus solely on factors that are intrinsic to the problems of fluid mechanics; he should attempt to reconstruct the state of the theory, the source and selection of experimental data, and the physical interpretation of the results. This done, his task is finished.

A second answer would be that a correct reconstruction of the problems of fluid mechanics requires concentration on external factors, such as the needs to which government institutions of the time gave priority and the philosophical and cultural stimuli that characterized the environment in which the monograph was produced.

Finally, a third answer comes from those who hold that internal and external history are complementary and should therefore be melded into the folds of one comprehensive history.

All three answers share a common problem: What are the reasonable criteria one should follow in drawing a boundary line between internal and external history? To what extent is the enunciation of a theorem part of internal history? And to what extent are one's philosophical tenets as to what is and what is not scientific part of external history? This problem is, of course, far from irrelevant. On the contrary, it and its possible solutions play a significant role in what is meant by a rational reconstruction of history.

There is another, more general, problem all three types of history must face: What is the relationship between theory and experiment? The history of science is in fact the history of the relationship between theories and experiments and so can never be reduced to a simple chronological exposition of theoretical and observational events carefully divided into two distinct categories. The interweaving of these two strands, however, is the source of even greater problems.

In essence, the problem is that nature does not give any answers to those who stare at it, or experience fragments of it, without being guided by expectations. In order to force nature into answering us we must act on it by posing questions; we must, that is, design experiments, manipulate techniques and machinery, propose hypotheses and theories, and interpret answering signals given in symbols that are not always clear or unequivocal. The art of interrogating nature has been singularly enriched in the last few centuries, and today the so-called natural phenomena do not have much to say to those who study them in poorness of thought. It is not due to chance, then, or to philosophy's malice, that the historical reconstruction of the art of scientific inquiry is increasingly concerned with the question of the relationship between theory and experiment.

Formulating the question in these terms may cause some difficulties. Who could claim, for instance, that the job of the theoretical physicist is merely to "translate" into symbols the knowledge gathered and ordered by the experimental physicist? Would anyone say that Kepler's laborious calculations were just a lifelong attempt to translate the empirical laws of planetary orbits into formulas? Doesn't the idea of translation really imply far too simplistic a representation of theoretical and experimental research?

We do not avoid the danger of oversimplification by claiming that the work of the theoretical physicist is purely inventive and that his results are unfettered products of the mind. Only a romantic would dare portray Newton and Dirac as solitary geniuses waiting for the blinding flash of revelation to strike them. The fact is that even in its moments of greatest abstraction theoretical physics retains a complex network of ties and connections with what can be experimentally tested.

We can proceed farther into our subject—the evaluation of the relationship between theory and experiment in its historical context—by trying to refine the idea that theorizing is in some way equivalent to translating. Let us take as an example one of the most important translations that physics has produced in the post-Newtonian era: the formulation of the electromagnetic theory of light.

One usually learns that this theory was developed during the last century in two distinct phases. During the first phase some scientists, Faraday in particular, patiently gathered and systematically cataloged a large body of data on electric, magnetic, electrochemical, and optical phenomena. During the second phase other scientists, Maxwell in particular, translated this empirical knowledge into mathematical formulas. The success of the translation is interpreted as having been due to the new theory's ability to condense this body of knowledge into concise and rigorous formulas.

At first sight this seems to have been a peaceful revolution, one based on faithful adherence to the rules that any honest professional, in this case a man undeniably possessed of a remarkable intelligence, is supposed to follow. In other words, Maxwell simply made correct use of a body of rules that could establish a correspondence between the laws discovered by Faraday in the laboratory and a mathematical language already available in texts.

By this hypothesis, the theorist Maxwell, guided by a mechanistic view of the world and of science, translated into precise equations a body of statements about facts that the experimentalist Faraday, not being himself a mathematician, had already written in a nonformalized and therefore somewhat ambiguous language. This sounds reasonable, but when we draw the logical consequences of this hypothesis we find several unpleasant consequences.

To translate one language into another one needs a dictionary. Assuming for the moment that we know Maxwell's dictionary, let us compare the conception of electromagnetic field proposed by Faraday (without mathematics) and the theory formulated by Maxwell (with mathematics and the dictionary). The first problem we find is that in Faraday's conception there is an implied interaction between electromagnetic and gravitational fields.[2] In Maxwell's conception, on the other hand, the interaction between electromagnetic and gravitational fields engenders paradoxes having to do with energy and admitting of no solution.[3] According to Faraday, therefore, a field theory should take into account such interaction, while according to Maxwell a field theory should eliminate it.

Here, then, is the first unpleasant consequence of the translation hypothesis: the translator, with due reference to his dictionary, unequivocally states that a significant part of the propositions to be translated cannot in fact be translated without incurring insoluble paradoxes. But this is a strange and worrisome translation—strange in that it casts doubt on the validity of the document to be translated, and worrisome in that this doubt is resolved by eliminating a portion of the document. This means that in the act of translating Maxwell has eliminated some fundamental elements of Faraday's physics.

Following this translation, as it happens, some seemingly elementary problems of optics cease to be elementary and appear instead quite complex. In order to solve the new problems some not insignificant portions of the translation must be redone. But instead of achieving a satisfactory solution this revision raises additional and even more serious problems of electrodynamics. This is another unpleasant consequence: it seems that in the process of translating Maxwell has added new difficulties to Faraday's physics.

To summarize: if the formulation of a theory is a translation into formulas of what is already known, the minimum preliminary guarantee one can demand of the translator is that he be faithful to the text, that is, to experience. But the result is invariably an unfaithful translation, which is a highly inconvenient and distressing turn of events. One might even conclude that theory is actually a systematic betrayal and that theorists are individuals who practice intellectual dishonesty while pretending to elaborate elegant mathematical correlations between phenomena.

If we wish to avoid these two unpleasant consequences, and the total condemnation of theory, there are several approaches open to us. The first is to defuse the problem. Those who favor this approach suggest that we should not give excessive weight to the differences that emerge during the process of translation, but should focus our attention, historical and philosophical, on Maxwell's mechanistic program. Given this premise, we may legitimately ask such questions as the following: To what extent is Maxwell's physics a mechanistic physics? Or, To what extent does Maxwell obey the tenets of Newtonianism? Defusing the problem in this case means reducing it to banality. There are hundreds of answers to the question of how much Newtonianism or mechanism is present in Maxwell. Furthermore, there are even answers that would equate Newtonianism with mechanism. None of these answers, however, is capable of breaking out of the metaphors of a second-rate philosophy, in the sense that none of them can tell us anything about the actual process of formulation of Maxwell's theory. This approach of systematic banality simply resorts to the artifice of writing the history of Maxwell's method in the main body of text and relegating the history of Maxwell's equations to the footnotes at the bottom of the page. In other words, it uses a variant of external history that focuses on the relationships between method and society.

The second approach starts from our original assumption that the formulation of a theory (Maxwell's) should be a faithful translation of empirical knowledge (Faraday's) but gives due consideration to the conclusion we have been forced to admit that a faithful translation does not exist. To explore this approach our next step is to consider the possibility that the problems encountered on the way might stem from the very nature of the dictionary used in translating Faraday's physics into Maxwell's physics.

Strong and Weak Interactions in the Dictionaries

We will now attempt to define the dictionary whose existence we have previously taken for granted. A study of Maxwell's strictly physico-mathematical papers shows that his dictionary was quite extensive and comprised an aggregate of theories: hydrodynamics, both Lagrange's and Hamilton's mechanics, the body of electric and magnetic theories of the French and German schools, several branches of mathematics, and some relevant chapters of astrophysics. The list goes on. Maxwell also studied logic and had a professional interest in the history of physics. Scattered throughout his various manuscripts we find important philosophical remarks on the function of models and analogies in physics. He upheld some concepts of molecular dynamics by invoking theological texts and carried out pioneering experimental research.

Thus, Maxwell's dictionary is not only extensive and complex; it is also extensively cross-referenced. Maxwell's researches on Henry Cavendish and eighteenth-century electrology are not unrelated to his view of the relationship between experiment and theory in physics; nor are these researches free from references to his laborious comparative analysis of Faraday's work and the theories of action at a distance expounded by Poisson and by the German school of physics. Maxwell's theses on statistical mechanics, unlike those proposed by Boltzmann in the same years, deal with both the cognitive power of the calculus of probability and the impropriety of advancing a fundamental hypothesis concerning molecular chaos inconsistent with the divine creation of molecules.

What we see here is an intricate network of correlations crossing the various levels into which the dictionary is subdivided. Not all these correlations have the same strength. The interactions between Lagrange's mechanics and the problem of models are relatively strong. Relatively weak are the connections that link, through a series of inferences that can be historically traced by successive approximations, the general principles of statistical mechanics of 1879[4] to "A Discourse on Molecules" of 1873,[5] wherein some arguments of a theological nature are taken to play an explicit role in the theory of knowledge.

Consequently, the task of the historian of science is to analyze the various levels of the dictionary and to evaluate critically the correlations that form its inner framework. It is by no means necessary—on the contrary, it is actually dangerous—to attempt to establish, once and for all, a demarcation line between "external" and "internal" that would split the dictionary in two even before we have had a chance to study it, or to discriminate on philosophical grounds in favor of some correlations and against others. The demarcation line is a myth of internal history, and the discrimination in favor of certain social, economic, and psychological factors is the delusion of external history.

Thus far we have spoken of dictionaries as solid structures. But a dictionary is not only vast, complex, and interwoven with correlations of varying strength; it is also unstable. Its components are subject to changes in time, and the rates of change can differ greatly from component to component. In other words, we are faced with a dynamic situation whose conformation is historically given, in the sense that we can point to specific areas of the dictionary that change with time and to correlations that vary in intensity.

In the case of Maxwell, for instance, examination of his publications, manuscripts, and private correspondence reveals that over the years large areas of his dictionary underwent drastic revisions. We need only observe the change in Maxwell's conception of the electromagnetic field from 1854 to 1873 to convince ourselves of the following fact: Maxwell's theoretical physics—his field theory and kinetic theory of gases—underwent a set of readjustments, a progression of deductive chains involving a number of theories that were rapidly evolving in response to the work of other physicists and to a stream of new ideas.

This shows that the dynamics of instability leads to a wealth of unsynchronized facets. Certain areas of the dictionary evolve more rapidly than others, and this implies that the correlations between areas also change in form and strength.

As we remarked above, it is the historian who attempts to retrace these levels and correlations. But this is not enough. He must also follow the dynamics of the evolutionary process for both levels and correlations. In short, the historian must do a lot of work if he wants to uncover a pattern. More important, there is something he positively must not do, namely, delude himself and others that it is his duty to extract from a scientist's dictionary only the scientist's method. The dictionary is vast, complex, deeply interconnected, and unstable, and its parts are out of phase. If we try to render it limited, simple, stable, and homogeneous by distilling its methodological essence through the alembics of some ready-made philosophy, we will end up with a caricature.

Maxwell's physics cannot be derived from Maxwell's method because his physics is neither a corollary of hypothetical theorems on method nor a logical consequence of decisions taken on philosophical grounds for the purpose of defending Newtonian laws at any cost.

We must insist on this point. Any historian who sought to uncover the structure—perhaps profound!—of Maxwell's science by using a label as well known and as much abused as mechanism would certainly get lost in myth. Between the tenets of mechanistic philosophy and Maxwell's equations there is no direct chain of reasoning but rather the whole of Maxwell's dictionary and the entire history of evolution of that dictionary.

Whoever embarks on a metaphorical journey through so-called nine-teenth-century mechanism will encounter not theories and scientists but philosophical allegories. Back from such a journey, in the privacy of his study, he will only be able to draw imaginary maps and will never carry out a single real exercise on the history of nineteenth-century physics. Led by illusory signposts, he will follow a mythological itinerary and will narrate dreams. Nor will he be saved from the delusions of a false history by the rules of some sophisticated and subtle traditionalist doctrine. The scientists whose dictionaries he proposes to explore were not engaged in a game of free choice among different concepts and postulates. They were engaged rather in building ideas and theories, and they could not choose before building—even though nobody can build objects or theories well unless he has knowledge of what has gone before him.

Several things, then, occur in a dictionary. The interactions between levels are a function of time. A modification in one area may cause slight variations in another area or radical alterations at seemingly remote levels. In sum, we can assume that the phenomena occurring in a dictionary do not have linear developments, cannot be explained by a method of trial-and-error, and may not be due to strict rules of cause and effect.

Furthermore, a dictionary is an open process: it interacts with other dictionaries, it may at times absorb some of their relevant chapters, and it is subject to local readjustments in an effort to reach a situation of relative stability. In Maxwell's case the dictionary is not only the dynamic matrix within which he builds a theory that describes the world rationally; it is also an element of the historical development of the quest for knowledge in the second half of the nineteenth century.

If all this has no part in Popper's logic or in the naive applications of classical causality, it does not mean that it can be reduced to irrationality. What it means, instead, is that the historical reconstruction of nonlinear processes and the study of ever more complex logics are wide-open, fertile fields.

It is a legitimate assertion that there are substantial differences between Einstein and an amoeba and that the whole of these differences constitutes history.

2 *The Galileo–Dirac Proposition*

Galileo's courageous battles have not necessarily bequeathed to us a secure legacy. It is a well-known fact that legacies are often a source of dispute and litigation, particularly when the beneficiaries are many, and some of them untrustworthy. All too often, in the process of infighting among the heirs the legacy is dissipated.

From this point of view Galileo's case is truly exemplary. The endless philosophical trial of Galilean science coincides with periodic attempts, planned and carried out by certain philosophical factions, to invade the realm of scientific thought. Tenaciously reopened by modern "conventionalism,"[1] this trial has culminated of late in the denigration of experience[2] and in the hypothetical graces of a qualitative dialectic that dreams of a world without science.[3]

Thus it is not surprising to find people indulging in the fanciful notion that schools ought to teach method rather than theories, as though method were everlasting (whether one wishes to condemn it or to exalt it) and theories destined to perish. This fancy germinates from the seeds of a classical misconception of anti-Galilean philosophy, which may be summarized as follows: in the first place, scientific method is a body of critical rules that enables us to formulate theories; in the second, theories, unlike method, are short-lived in the history of scientific thought; in the third place, the relative longevity of method (Galilean) is due to its philosophical foundations; finally, and in conclusion, the most relevant and important portions of scientific thought—those portions that endure—are essentially philosophical. Thereby individual theories get buried in the graveyards of history or are portrayed as mere provisional techniques.

We are faced here with a general criterion of demarcation that would divide man's mind and his culture into two separate domains: the scientific and the philosophical. Once we accept that a criterion of demarcation exists, we open the door to the catechism of those who reduce either science to philosophy or philosophy to science. And in both cases the roots of one doctrine are sought in the terrain of the other. The contraposition between internal and external history originates precisely from this fallacy and from the consequent impulse to search for the roots of scientific thought in philosophical thought or vice versa.

The unity of knowledge, on the other hand, is the basic assumption in the work of the historian who analyzes the dictionaries, studies their structure, and reconstructs their development. From all we suggested in the preceding chapter it seems rather plausible that a dictionary is neither a dogma one should adhere to on faith or for propagandistic reasons, nor the nucleus of a research program that one or more scientists institute by means of a methodological fiat and obstinately defend from every attack. The historical analysis of the dictionary does not put undue emphasis on the motivations of a scientist who, faced with a crisis, gives up a cherished model with arguments not unlike those of a betrayed lover or a thief caught red-handed; nor does it favor the whims of the methodological faction.

In the study of the dictionary the question of crises is bound to arise. But we are dealing with local crises, since they involve only specific levels of the dictionary that are particularly sensitive to the readjustments taking place at other levels. There is nothing strange in the fact that Poincaré's theorems on hydrodynamics (1892) should have forced the great Kelvin to abandon his theory of the structure of matter, and that Kelvin should have considered this necessity as an actual failure (1896).[4] But it would be strange and untenable to claim that Poincaré's theorems provoked a crisis and a failure in physics, and consequently created psychological conditions favorable to the rise of a new paradigm. Those who would have it so must pay the not negligible price of confusing a revolution with a crisis.

The study of the dictionary also dispels the philosophical delusions of those who claim that "a brilliant school of scholars (backed by a rich society to finance a few well-planned tests) might succeed in pushing any fantastic programme ahead or, alternatively, if so inclined, in overthrowing any arbitrarily chosen pillar of 'established knowledge.' "[5] Such things have certainly happened, inasmuch as there is also a science entrapped by dogmas and, perhaps, well paid. But the fact that these things happen must not be justified on rational grounds unless we truly wish the planning of research to hinge on a conservative, obscurantist policy.

If, as a result of rational reconstructions that place real history in the footnotes and internal history in the main body of the text, one is brought to praise a "human creative imagination"[6] nourished by a rich society, then it is clear that the history written by a rigid internist is no different from the history written by a rigid externist. In this coincidence of seemingly opposite positions the dissolution of empiricism is consummated and the eclipse of objective knowledge appears to triumph.

By going beyond the pseudoseparation between internal and external history, we recognize that the theory of knowledge of a certain physicist (or mathematician or chemist) is not defined by the sum of his methodological rules but is given by the whole of the correlations between those rules and all the theories and experiments the scientist has conceived over the years. A theory of knowledge is not a philosophy but a historical process.

All this is so true that—apart from the fact that Galileo never wrote a *Treatise on Method*—the law of fall of bodies could never be derived from the Galilean method. The correlations between methodological rules and physical laws are not syllogisms; rather, they are interactions, changing with time, between variable laws and rules. Rules, in fact, are reflections on the laws, and the laws take into account the fact that nature does not always answer our questions in predictable ways. Theories do not grow because nature is an onion whose layers science is peeling away. On the contrary, science lives in the constant need to reinterpret knowledge already acquired as soon as a deeper level of reality comes under investigation. It is not a process of accretion, then, but a constant restructuring of theories and methods.

It was nature's unpredictable answers that exposed the excessive pretensions underlying Galileo's thought, proudly stated in his assertion that our knowledge of the world is indeed limited to certain areas but that within those areas it equals that of god.[7]

In other words, the strangeness of nature has taught us that there are no absolute and unalterable forms of scientific knowledge. It has also taught us that knowledge grows thanks to "theories that violate the senses" while availing themselves of "sensible experiences" and "exact demonstrations." For us today, Galileo is still arguing with Simplicius that mathematics is the guarantee of reason (in those areas of the dictionary where it operates). The interaction of sensible experiments and exact demonstrations is a historical process; it does not lead, therefore, to confrontations to be decided in the tribunals of classical logic or in the praetorial chambers of crucial experiments.

For these reasons we can still uphold the Galilean thesis, which we can summarize as follows: (a) "When the philosopher-geometer wishes to recognize in the concrete world the effects demonstrated in the abstract he must cut down the encumbrances of matter";[8] (b) "Our discourses must relate to the sensible world and not to one on paper."[9]

It is also true, however, that Galileo's thesis must be amended, for the history of the post-Galilean physical sciences proves that theories reflect on themselves and on the theories that precede them: objective problems do arise in the paper world. And the amendment recognizes the growing abstraction of theories.

Around the middle of the nineteenth century the great mathematician Riemann defended the profound rationality of a knowledge that progressively detaches itself from the raw data of the senses and of everyday intuition: "Little by little our conception of nature becomes increasingly more complete and exact, but at the same time further and further removed from superficial appearances."[10] And it was the mathematical physicist Boltzmann who denounced the absurdity of regarding theory as "a milk cow," or a mere technical instrument, and who made an impassioned plea for abstract thinking: "The more abstract the theoretical investigation, the more powerful it becomes" and conquers the world.[11]

Some philosophers perhaps will be irritated by the fact that the problem of objectivity is formulated in terms that may be reminiscent of the battles between the "old" materialism and the "old" rationalism. All too often, in fact, the perception of the technological constraints conditioning contemporary man generates neoromantic rejections of reason and its triumphs. An alarming symptom of this trend is the facility with which the term *scientism* is used, oftentimes inappropriately, to criticize the position of those who consider scientific knowledge as the best way to perceive nature.

The liberalization of the Galilean theses—as it emerges from the thought of a Riemann or a Boltzmann—and the growing power of theory over the world have encountered moments of subtle tension in contemporary physics. Even in the limited context of these reflections it may be appropriate to recall a page written by Dirac in 1931:

The steady progress of physics requires for its theoretical formulation a mathematics that gets continually more advanced. This is only natural and to be expected. What, however, was not expected by the scientific workers of the last century was the particular form that the line of advancement of the mathematics would take, namely, it was expected that the mathematics would get more and more complicated, but would rest on a permanent basis of axioms and definitions, while actually the modern physical developments have required a mathematics that continually shifts its foundations and gets more abstract. Non-euclidean geometry and non-commutative algebra, which were at one time considered to be purely fictions of the mind and pastimes for local thinkers, have now been found to be very necessary for the description of general facts of the physical world. It seems likely that this process of increasing abstraction will continue in the future and that advance in physics is to be associated with a continual modification and generalization of the axioms at the base of the mathematics rather than with a logical development of any one mathematical scheme on a fixed foundation.[12]

Retracing the steps in the liberalization of the themes of Galilean thought —a liberalization required by the discovery and investigation of new worlds by the physical sciences—we realize that the various branches of mathematics and physics develop over the decades and grow not on eternal pillars but on foundations susceptible to change. One of the chief attributes of what has been defined as the second scientific revolution is precisely the recognition of this basic property of objective knowledge. The second scientific revolution does not consist merely of a long and complicated process of restructuring of the physical laws; it also entails an awareness of the depth of such restructuring and of its historical development. It is precisely on the basis of this awareness that we can judge the philosophical dimensions of the second scientific revolution.

From what has been said up to now we can draw, therefore, the following conclusion. The study of the dictionaries enables us to evaluate the magnitude of the errors intrinsic in the distinction between internal and external history of science, in the sense that such errors are found to stem from improper notions concerning the problem of objective knowledge. Hence we can suggest that objectivity does not reside in a specific method but in the historical process that causes the various methods and the different forms of explanation-by-rules to vary as a function of the answers that nature gives to sensible questions. This implies that objective knowledge, insofar as it sends man back to nature as the mother of necessity, is predicated on a materialistic axiom. To be correctly enunciated, this axiom must present itself to reason in the historically given forms of what we can call the Galileo–Dirac proposition.

According to this point of view, science never grows in ivory towers, for it is never unaffected by the weak interactions that establish relationships, within the dictionaries, between theories and philosophical conceptions of the world. At the same time the construction of individual theories is relatively autonomous with respect to philosophical views of the world since it is entrusted to locally stable areas of rules within the dictionary and to the answers, often surprising, that nature gives to sensible experiments. And since the variations that affect the rules and modify the strong interactions between such rules and theories are also subject to locally stable criteria— theories of demonstration, analyses of the logic of theories, and so on—it follows that scientific enterprise, if it does not have absolute foundations or categories, nonetheless is not without objective bases that can be rigorously tested.

A significant characteristic of the dictionaries is therefore the following: in view of their complexity and historical dynamics, it is well-nigh impossible to adhere to them for reasons of faith, or to defend them by means of methodological decrees. It is certainly possible to adhere dogmatically to certain areas of the dictionary, or to choose to defend just as dogmatically certain other areas. In other words, "spontaneous philosophies" are actually practiced in science, and they are relevant to the extent that they motivate scientists to prefer some areas of rules to others, or to look for some justifications rather than others. We need only recall the qualitative dialectic at work in Bohr's thought or the naive realism coloring Einstein's view of the world. But the fact that we can reconstruct some aspects of the spontaneous philosophy of a physicist or a mathematician by no means implies that we should resolve his physical or mathematical theories into that philosophy. We are dealing with problems concerning the interactions between a dictionary's levels, not with problems concerning the reduction of one level to another, or the elimination of certain levels particularly ill suited to rational reconstruction. While the intransigent internist eliminates specific areas of a dictionary on the ground that they are irrational, and the intransigent externist turns history into ideology, the historian of dictionaries endeavors to put into practice the thesis that science does not develop by the shortest logical path.

II
STUDIES ON THE SECOND SCIENTIFIC REVOLUTION

Our journey into "classical" physics begins with the recollection of a controversy that breaks out in 1886. The bone of contention is the relationship that should exist between mathematics and physics. Is mathematics the servant or the master of physics and experimental investigation? If physics comes to know the world thanks to the painstaking and repeated observation of facts, and if mathematics follows observation and is no more than a set of rules for writing out known laws in rigorous form, are we not forced to admit that mathematics is a mere tool of thought? And, on the other hand, is it not a justifiable claim that experiments cannot be devised without recourse to theories?

Our choice of a scientific view of the world depends on the answers we give to these questions, and there are conflicting answers.

The great Kelvin (hailed by his contemporaries as a second Newton) accuses—and has others accuse—Boltzmann of being a disciple of an anti-Newtonian philosophical sect. Boltzmann is charged with attempting to substitute mathematical deduction for physical thought, thus falling into heresy with regard to the precepts of empirical research.

The rules of empirical research, Kelvin proclaims, have a cornerstone that bears the inscription: Experience is the only source of knowledge. Clearly, prescriptions of a philosophical nature are at work in the dictionaries of physicists like Kelvin and P. G. Tait. In the light of such prescriptions, mathematical abstraction is incapable of directing research on the physical world, and its only value is in expressing what has been apprehended through the inductive method. Mathematics is indeed a splendid edifice, but its body of rules has no value unless it represents the idealization of common sense. Consequently, a sector of the English school of physics mobilizes against Boltzmann's theories, and, by adopting an epistemological optimism that emphasizes common sense and the "looking at things" before turning them into mathematics, it creates an image of nature and of science that claims to be securely anchored in common sense and in the tradition that has very nearly equated Bacon with Galileo, and Newton with Thomas Reid.

Research into the various levels of Kelvin's dictionary does not lead us to retrace the traditional pathways of mechanism. Rather, by following the interweaving of connections and complex interactions, it leads us to re-examine the influence that specific cultural traditions exert on the determination of what good science must be—and there is no doubt that Kelvin's physics was good physics. What poses problems, and far from trivial ones, is something else, namely, the fact that we no longer see classical physics as a homogeneous block of explanations accreting onto a common nucleus of methodological rules. And since all this bears on the function of models in relationship to theories, it is not excessive to claim that in choosing certain chapters in the book of mathematics rather than others, physicists are not just practicing a bit of so-called spontaneous philosophy but are actually determining the basic character of the theories they themselves are constructing.

3   *Herschel's Lion*

Mathematics and Experience

You substitute mathematics for thought and thus betray healthy empirical research. Instead of using mathematics as an aid to thought in the investigation of reality, you turn the physical sciences into an empty play of abstract symbols and hide thoughts behind a barrier of "terrific arrays of signs." This is the harsh indictment that was pronounced at the end of the last century against the body of theoretical studies developed by Ludwig Boltzmann, one of the greatest scientists of the modern era, in his attempt to understand and explain the laws of molecular motions.[1]

This indictment, it should be noted, was not directed against the validity of the hypotheses concerning molecules. Other scientists and philosophers were already doing that by asserting that conjectures about molecules and atoms were mere philosophical illusions for which no experimental confirmation could be found. The indictment we are discussing here consisted instead in denying that mathematics had any cognitive value. From the point of view of the accuser, mathematics was a tool whose proper use was as an aid to the thinking process. In this capacity it was an eminently precise and controllable language. As a language and a tool it made it possible to express known laws in rigorous form and to predict the consequences of those laws. But just because it was a language and a tool, it was unthinkable that it could somehow become a substitute for thought. The thought process used in the study of natural phenomena and events had rules of its own and derived its certainties from domains that were not those of mathematical abstraction. The latter, left to itself, was nothing but a play of symbols. And whoever aspired to think about nature and investigate its underlying order by relying on mathematical abstractions would end up with no more than a game of symbols since the source of man's knowledge of nature was something quite different from what appeared in the "terrific arrays of symbols" that formed the framework of Boltzmann's theories.

The indictment, in sum, implied that Boltzmann's theories were not physical theories, but mere mathematical constructions incapable of speaking sensibly about the world.[2] In the eyes of his critic, Boltzmann appeared as somebody who while climbing toward abstraction was losing sight of reality.

Let us find out, then, what is the proper method of constructing a healthy physics according to Boltzmann's critic. His name is Peter Guthrie Tait and he is considered one of the most important physicists of the second half of the nineteenth century. A pupil of the great mathematician William Rowan Hamilton and a collaborator of *the* Lord Kelvin who is acclaimed by the scientific world as a second Newton, Tait has a profound knowledge of the mathematical and physical sciences of his time. He is also a cultured person in the usual sense of the word, and is particularly attuned to the implications of the philosophical battle over science and religion that is raging in European culture. It is his firm belief that the good physicist must be a good Newtonian, and that to be a good Newtonian he must be able to draw a clear demarcation line between what pertains to physics and what pertains instead to other forms of inquiry. Only by taking this position can the student of natural phenomena keep his mind free from the "harmless folly" of the spiritualists and from the "pernicious nonsense of materialism." The validity of this position is guaranteed by a precise norm: experience is the only source of knowledge.

According to Tait, in the physical sciences "advances come or not according as we remember or forget that our science is to be based entirely upon experiment or mathematical deductions from experiment."[3]

If scientific knowledge is to grow healthy and strong, Tait writes, we must follow the rule that "there is nothing physical to be learned a priori." In essence, this is the basic premise of the fight against Boltzmann, whose mathematical terrorism conceals its cognitive impotence behind pages of formulas. As Tait remarks, "We have no right whatever to ascertain a single physical truth without seeking for it physically, unless it be a necessary consequence of other truths already acquired by experiment, in which case mathematical reasoning is alone requisite."

We now begin to understand the reason for the opposition to Boltzmann's theories: they claim, mistakenly, to acquire knowledge by means of rules that apply solely to abstract symbols and that, if useful in deducing logically necessary consequences, are not in themselves capable of coming to grips with reality. It is only after experience has given us the general laws binding the physical world that an abstract language becomes useful, since it enables us to derive the necessary consequences "of other truths already acquired by experiment."

What are the foundations of Tait's firm belief in experience as the only source of knowledge?

At its root, as our critic writes, is belief in the supremacy of empirical knowledge, a supremacy that is justified in itself and appears in the form of "a self-evident principle." Only the faithful followers of Newton's experimental philosophy can appreciate the truth of this self-evident principle and thus avoid being led astray by the error of those who attempt to investigate nature by means of mathematical deduction alone. At this point we might remark that Tait is completely blind to the enormous weakness of the thesis he advances—a weakness stemming from the fact that admittedly it is founded on an assumption that should be self-evident, that is, evident to anyone who has not fallen prey to some philosophical nonsense. As a matter of fact, the victims of such nonsense have already been identified and classified. As Tait himself says, they are the misguided people who pursue the follies of spiritualism and material-ism. In consequence, the minds of these people will never be illuminated by the light of self-evidence, which shines instead in the thoughts of those who are able to distinguish between science and everything else. The true Newtonian, Tait writes, is always able "to face the question, where to draw the line between that which is physical and that which is utterly beyond physics. And again, our answer is—Experience alone can tell us; for experience is our only possible guide. If we attend earnestly and honestly to its teaching, we shall never go far astray."[4]

The Tait–Boltzmann controversy is a very significant example of the contraposition of two different ways of perceiving physical knowledge of the world, particularly because this contraposition has a direct bearing on a number of questions concerning mathematics and experience. Moreover, we are not dealing with an example of limited relevance. Even if Tait were considered a minor scientist, this in itself would not be sufficient reason to underrate the case at hand. The relevance of the example is that Tait launched his attacks on Boltzmann's theories at the explicit urging of Kelvin, that is, following the directives of the most influential physicist of the time. This leads us to consider the view of the physical world, mathematics, and the value of experience that emerges from the investigations jointly carried out by Kelvin and Tait.

Mathematical Analysis and Phenomena

One of the best treatises on physics to appear in the last quarter of the nineteenth century bears the signatures of Kelvin and Tait.[5] The first of the two volumes that constitute *A Treatise on Natural Philosophy* surprisingly begins with a quotation taken from a celebrated mathematicophysical work by J. B. Fourier: "The primordial causes are unknown to us; but they are subject to simple and constant laws which one can discover through observation, and whose study is the object of natural philosophy."[6]

Why is this quotation surprising? Because it would be very difficult to misconstrue Fourier's thought in such a way as to present it as a methodology based on the denial of the cognitive power of mathematics. It is certainly true that in the first decades of the century Fourier had repeatedly advocated the necessity of assuming as premises to theories some general and simple propositions derived from observation rather than models. But it is equally true that Fourier never thought that the criteria of validity for physical theories were dependent on the confirmation of such theories by means of experiments. According to Fourier, the formulation of a physical theory must take place in accordance with the rules of calculus and independently of the physical interpretation of the terms that appear in the calculus. Only a theory endowed with internal coherence could reasonably lead, in a purely deductive way, to equations capable of expressing relationships between phenomena. In Fourier's methodology and scientific practice we see the triumph of mathematical reasoning rather than the glorification of the inductive method. According to Fourier, in other words, mathematical analysis and theoretical physics go hand in hand, and the validity of a physical theory is predicated on the norm that "mathematical analysis has necessary connections with physical phenomena; its product is not a creation of man's mind but an innate element of universal order which is neither transitory nor fortuitous: it is imprinted on all of nature."[7]

For Fourier, then, there is a peculiar relationship between mathematical abstraction and observation. The solutions of equations describing phenomena are necessarily valid if the deductions used in formulating them are correct. Furthermore, a scientist has the additional guarantee that the derived solutions consist of simple mathematical propositions, each of them revealing "an underlying phenomenon that becomes perceptible through experiments."[8]

To leave no doubts about the relationship "existing between the abstract science of numbers and natural causes," Fourier states that, on the one hand, "we cannot make any modification to the form of our equations without depriving them of their essential character, which is that of representing the phenomena,"[9] and, on the other hand, that when a mathematical solution is rewritten as a series of simpler functions, we have to admit that we are not confronted only with a necessary consequence of calculus. "The natural phenomenon whose laws we are searching for," Fourier writes, "is actually divided into distinct components which correspond to the various terms of the series."[10]

From this point of view, it is entirely tenable to consider theories as deductive structures capable of suggesting the whole of sensible experimental observations. In other words, the theory of heat phenomena is valid because it is mathematically true, and because there is a profound connection between what is mathematically true and natural entities. As a result, Fourier deems it quite appropriate to inscribe the motto "*et ignem regunt numeri*" at the beginning of his *Théorie analytique de la chaleur* of 1822.

Fourier's view of the relationship between mathematics and observation is not the only reason why it is so startling that Kelvin and Tait should quote one of his sentences. We must also remember that Fourier strongly denies any possibility of reducing his theory to the principles of mechanics.[11] Fourier is quite explicit on this point: "Whatever the scope of mechanical theories may be, they are not at all applicable to the effects of heat. The latter constitute a special order of phenomena that cannot be explained by the principles of motion and equilibrium."[12] If there are sectors of natural philosophy that cannot be referred to dynamic theories but have laws of their own,[13] it would not be reasonable to try to formulate an explanation of such phenomena by means of Newtonian principles, or to search for a solution in terms of models or hypotheses concerning the nature of heat itself. What is important is only the "knowledge of the mathematical laws" that govern thermal effects.[14]

Consequently, the theory of heat is an autonomous theory and its constituent elements, in Fourier's opinion, can be reduced to operations of exact definition of all the mathematical terms, of the differential equations and "the integrals relative to the fundamental problems."[15]

There is no doubt that Kelvin studied Fourier's mathematical and physical work in the greatest detail. Kelvin's early writings are a brilliant commentary on the *Théorie* of 1822 and a passionate defense of the mathematical methods developed in it. Thus when Kelvin and Tait pay tribute to Fourier in their 1879 *Treatise*, the fact that they quote one particular sentence from the *Théorie* that states that simple and constant laws are discovered by observation has a precise meaning. In the first place, it is out of the question that the use of Fourier's thought in support of empiricism could be due to a superficial analysis by the two English physicists of the work of the great French mathematical physicist. Tait's mathematical background and Kelvin's own study of Fourier make this view untenable. In the second place, we must remember that physics' involvement with philosophy is not simply the questionable and ambiguous subject of philosophical slogans. The fact that the sciences are not indifferent to philosophical thought forms the true fabric of scientific practice. In short, scientists think by using norms that are historically determined rather than by invoking fixed and unalterable canons.

The way the two English physicists interpret Fourier's axiomatic methodology is not very different, in certain respects, from the way Dutch scientists of the early eighteenth century interpreted Newton's mathematical glories. In both cases, in fact, the scientist's work was interpreted in the light of a form of empiricism. Newton's success thus depended in large measure on the fact that his theories were no longer accused of excessive mathematization and on the general acceptance of the view that his mechanics and optics were examples of inductive learning.[16] Similarly, the good fortune Fourier enjoyed among English physicists in the late nineteenth century had strong roots in an eclectic empiricism and in a realism we shall examine at length later on. For the moment we only wish to note that both sources had a determining influence on Kelvin and Tait's interpretation.

One point needs to be emphasized: whereas Fourier maintained that simple and constant laws are discovered by observation and that general laws concerning phenomena can only be discovered by pure mathematics, Kelvin and Tait saw Fourier as somebody who identified the simple and constant laws with the general laws. The success of this empirical interpretation is nothing but a philosophical metaphor in which the anti-Laplacian Fourier disappears to be reborn as an imaginary thinker, as one of those followers of Bacon who, according to Kelvin and Tait, fill the ranks of the good Newtonians. A metaphor, we said—but we should add that this metaphor affects history in a concrete manner, as an actual epistemological weapon. Not only does this weapon erase the differences that Fourier had pointed out between simple, constant laws and general laws concerning phenomena; not only does it ignore the fact that if the former concern what is perceptible to the senses, the latter concern instead what exists in the objective world and, as general laws, can only be discovered by deduction; most important, it demonstrates that philosophical speculation about the sciences does not live apart from the sciences themselves.

Philosophical speculation about the sciences is indeed an integral part of the historical forms that the sciences themselves assume in their dynamics, in the sense that their dynamics cannot be understood unless we take into account the boundary conditions imposed on it by philosophy in its role as an interpretative medium.

A Mathematical Terrorist?

The particular aspect of the scientific and philosophical situation we have just discussed enables us to understand the real meaning of Tait's accusations against Boltzmann and against theories in which mathematics seems to have become a substitute for thought. However, the interpretation of Fourier in an empirical key is only one aspect of that situation. Fourier, first and foremost, developed a generalization of mathematical methods. Once Fourier has been reinterpreted so as to emphasize experience as the only source of knowledge and to relegate mathematics to a later stage in the formulation of general laws by the inductive method, there still remains the problem of how mathematics should behave "after" the discovery of physical laws.

The *Treatise* of Kelvin and Tait is quite sensitive to this problem. Mathematics must be an "aid" to thought, not a substitute for it. But how can one comply with this norm if deductions obtained by mathematical analysis sometimes lead to unexpected predictions and to the identification of objects never previously observed? And why is it that in other instances mathematical deduction is used to prove the existence of objects that subsequently disappear from the scientific domain?

These are momentous questions for nineteenth-century physics. Consider caloric, for instance. The existence of this fluid was the cornerstone of the theories of thermal phenomena formulated by Laplace and his school during the first decades of the century. In the context of these theories caloric was an object in the full sense of the word.[17] Laplace's mathematics not only elaborated the properties of this object but allowed for the possibility of obtaining experimental proof of its existence in physical bodies. Subsequently, however, in the light of more advanced physical knowledge of heat phenomena, caloric was declared nonexistent. Did this fact necessarily mean that the objects discussed in physical theories were intellectual constructs rather than real "things"?

One way to answer this question was to advance the view that caloric was an object solely within a theoretical model of molecular structure. Thus its disappearance as an object did not trouble overmuch the physicists of the realist school, who could simply ascribe it to Laplace's questionable hypotheses and their downfall. On the contrary, the disappearance of caloric was a tribute to observation, since only observation had been able, through crucial experiments, to disprove conclusively all theories based on it. A physicist like Kelvin had no doubts on the matter. False theories had been overthrown by healthy experiments, and this confirmed that the only sure base for knowledge was to be found in experience.

However, there was no simple explanation for the fact that theories could predict what had not yet been observed and show the path along which the exploration of the unknown could be fruitful. Only telescopes and a patient observational technique had allowed John F. W. Herschel to discover more than five hundred new nebulas, but mathematics alone had enabled John Couch Adams and Urbain Le Verrier to predict the existence of the planet Neptune starting from a mathematical analysis of the perturbations of Uranus. On the basis of the indications communicated by Le Verrier the new planet was later discovered with the instruments of the Berlin Observatory on the night of September 23, 1846.

When deductive methods yielded such results, how could experience still be upheld as the *only* source of knowledge?

One could not simply dismiss the question by pointing out that deduction had had to start from observation of the perturbations in Uranus's motion. Observation alone, although painstaking and detailed, had not been able to find the new planet. Mathematics had taken its place and provided the necessary information. This fact was extremely important, and in his opening address at the meeting of the British Association held at Oxford in 1847, Robert Harry Inglis remarked, "The progress of astronomy during the last year has been distinguished by a discovery the most remarkable, perhaps, ever made as the result of pure intellect exercised *before* observation, and determining *without* observation the existence and force of a planet; which existence and which force were subsequently verified *by* observation." And in this achievement of pure intellect, Inglis added, the usual process of research has been reversed.[18]

In their *Treatise* Kelvin and Tait confront this question in a general way. Knowledge of planetary motions, they maintain, is founded on solid bases. Its foundations, in fact, are principles or laws that have been derived directly from experience: the laws of motion and the law of gravitation. It is true that these laws hinge on the concept of force. However, this concept is but "a direct object of sense; probably of all our senses, and certainly of the 'muscular sense.' "[19] Hence there is no paradox in the fact that in some instances one anticipates experience by following the path of mathematical deduction. Natural philosophy, Kelvin and Tait write, consists both of the empirical study of the laws of the physical world and of the mathematical deduction of specific results that have not yet been investigated observationally. The entire scientific enterprise is nothing but an "extension of our knowledge." When the forces are known, as in the case of planetary motions, "the mathematical theory is absolutely true, and requires only analysis to work out its remotest details." Thus, one of the obvious tasks of mathematical theory is to investigate the consequences of the laws. And while the laws are derived from facts, the "details" are deduced by analysis. There is nothing disturbing about the power of mathematics to uncover details. "It is thus, in general, far ahead of observation, and is competent to predict effects not yet even observed—as, for instance, Lunar Inequalities due to the action of Venus upon the Earth . . . to which no amount of observation, unaided by theory, would ever have enabled us to assign the true cause."[20]

Let us be careful, however. This acknowledgment of theory's role by no means implies a weakening of the privileged position of experience. The validity of mathematical theory stems from the certainty of empirical principles—the laws of dynamics and Newton's law of gravitational interaction—but it should not mislead us. In effect, the world is so complex that it is not possible for man to arrive at "the true and complete solution of any physical question by the only perfect method," that is, by the method that implies the knowledge of every form of motion. The absolute validity of a theory is ever dependent on the constraints imposed on it by a natural world that is enormously complex and extended in both time and space. The only approach a Newtonian physicist can follow is to appeal to nature's dictates, which dictates impose limitations. Thus the true Newtonian physicist must recognize the necessity of developing knowledge through a process of progressive extensions starting from principles discovered by induction—a process that has no end and whose stages are given by increasingly better approximations. This extension by successive approximations certainly requires mathematical methods. But, Kelvin and Tait warn, they must be used under precise conditions. In the final analysis, these conditions are based on the ever-present need to limit the excessive generality of the problems that can be formulated by applying mathematics to physics.

Consequently, certain stratagems are needed in the mathematical approach that will prevent research from getting lost in excessively general and formal questions. In order to keep research from becoming too abstract and too general, the good Newtonian will therefore make use of what Kelvin and Tait call "practically sufficient methods." Such methods avail themselves of "assumptions more and more nearly coincident with observation" and in addition bear the earmark of what the authors of the *Treatise* view as an essential criterion: "The limitations introduced being themselves deduced from experience, and being therefore Nature's own solution."[21]

There exist criteria, then, that can determine what mathematics must do "after" the empirical sciences have derived the laws of nature from the observed phenomena. And these criteria seem to be valid even for specifying which types of mathematical methods are useful to physicists. As we have seen, these methods must be "practically" sufficient. In this case practical sufficiency is not to be regarded as a consequence of the researchers' weakness but as a reflection of nature itself. In short, it is nature that prescribes the truly "natural" solutions to the good Newtonian and leads science to those extensions by successive approximations that form the genuine content of rational knowledge.

The research rules that we glean from the pages of the *Treatise* do not consistently favor experience over theoretical work. These norms grant theory a legitimate role while trying to prevent the damage that, in the authors' opinion, results from adopting too general an abstract method.

From the point of view sanctioned by these rules there can be no doubts about Boltzmann's methodology. The latter does not seem to accept the constraints directly imposed by nature, and appears as an attempt to substitute for empirical research the insane pretension to deduce natural laws a priori, using assumptions that are not nearly coincident with observation, as well as formal methods that are quite far from being practically sufficient.

The meaning of Tait's accusations against Boltzmann is now much better defined: Boltzmann is a mathematical terrorist precisely because he is not a Newtonian physicist.

The Senses and the Structure of Space: Tait and Riemann

The arguments advanced by Kelvin and Tait rest on a set of assumptions concerning matter. In their *Treatise*, for instance, we read that it is not possible to formulate "a definition of Matter" that would be considered satisfactory by metaphysicians. However, a definition can be formulated that will satisfy a scientific investigator. In the authors' opinion "the naturalist may be content to know matter as *that which can be perceived by the senses*, or as *that which can be acted upon by, or can exert, force*."[22] Which brings us back to the senses since, as we have already seen, Kelvin and Tait understand forces as objects of the senses rather than as intellectual constructs.

The study of matter is carried out primarily by observation and experimentation. The former is concerned with ascertaining causes "by simply watching their effects," as in the case of astronomy; the latter, instead, refers to what happens in the laboratory, where "we interfere arbitrarily with the causes or circumstances of a phenomenon."[23]

This distinction between observing and experimenting clearly reveals the origin of Kelvin's and Tait's methodology. As we shall see later on, their methodology derives in large measure from the concepts formulated by Herschel in 1830 and, through various intermediaries, from the common sense philosophy of the second half of the eighteenth century. One consequence of this doctrine that concerns us now is that with all his observations, experiments, and mathematical predictions, the Newtonian scientist of Kelvin and Tait can never fully know the world, even though he has at his disposal firm criteria for building such knowledge.

In a view of science in which experience is the only source of knowledge and mathematics merely a tool to derive the details, the objective world that confronts the physicist is infinitely knowable by successively better approximations; but the only way to ensure the objectivity of such approximations is for the physicist to look at nature through crucial experiments. The objectivity of the physical sciences, to repeat it once more, never rests on mathematics. Not by chance does Kelvin take every opportunity to depict mathematics in the guise of a servant and faithful helper and to refute the idea that mathematics is the master.[24]

In what sense do observation and experimentation touch upon an inexhaustible material world? The *Treatise* partly answers this question when it mentions the infinite complexity of nature. But we should not be deceived by these statements about an infinite complexity. Both Kelvin and Tait are convinced that such complexity can be gradually resolved by cognitive acts capable of building a progressive series of increasingly better approximations. The natural sciences advance by extending the laws found by induction and therefore never venture on an ocean of mysteries that are insoluble a priori.

According to Tait, the notion that nature is inexhaustible and science's work historically endless is a consequence of the fact that nature's ultimate constituent elements can never be identified. In 1876 the starting point of Tait's reflections is the realization that in the physical sciences "there is no such thing as absolute size."[25] Thus it does not make sense to maintain, for instance, that an object that appears exceedingly small through the most powerful microscope is of necessity an object devoid of inner structure. Nothing prevents that object, Tait writes, from being "astoundingly complex in its structure," and nothing prevents its inner layers from being as numerous and varied as those of a star.

Nothing is more preposterously unscientific than to assert (as is constantly done by the quasi-scientific writers of the present day) that with the utmost strides attempted by science we should necessarily be sensibly nearer to a conception of the ultimate nature of matter. Only sheer ignorance could assert that there is any limit to the amount of information which human beings may in time acquire of the constitution of matter. However far we may manage to go, there will still appear before us something further to be assailed. The small separate particles of a gas are each, no doubt, less complex in structure than the whole visible universe; but the comparison is a comparison of *two infinities*.

For Tait, the fact that the physical world is inexhaustible, in that empirical research can probe ever more deeply into each of its layers, has a clear implication, namely, the need to oppose those insane philosophies that claim to discuss the "so-called ultimate constitution of matter" and to prevent them from ruining the Newtonians' healthy physics with the mirage of ultimate and definitive elements that reason can grasp, and, at the same time, the need to fight the skeptics who would turn the endlessness of the learning process into a failure of scientific thought. We should note in this respect that no doubt can ever shake Tait's faith in experience as a scientific guide. Physics may not be able to give us ultimate knowledge, but this by no means implies that we should renounce a knowledge that grows on sound foundations. In the same decade, however, very different voices are heard in scientific circles. On August 14, 1872, at a meeting of German naturalists and physicians held at Leipzig, Emil Du Bois-Reymond declares that there are insuperable limitations to human knowledge. These limitations, he warns, are inherent in the very method of Laplacian mathematical physics and produce insoluble enigmas. Confronted with certain problems, the natural scientist can say *ignoramus* and at the same time harbor the hope that in a more or less distant future he will be able to understand. "But confronted with the enigma of what matter and force are and how they can be capable of thought," Du Bois-Reymond concludes, "he must bow once and for all to the verdict which is much harsher in its hopelessness: *ignorabimus.*"[26]

The clash between Du Bois-Reymond's ultimate renunciation and Tait's endless growth of empirical knowledge is quite evident. And Du Bois-Reymond does not refrain from harshly attacking the Scottish physicist. In 1880 he accuses Tait of harboring illusions and of not being able to resolve "the contradictions our mind runs into in the effort to comprehend matter and force." In his celebrated essay on "the seven enigmas of the world" Tait appears as "that professor Tait who with his chauvinism has rekindled the controversy on the part played by Leibniz in the discovery of infinitesimal calculus; who has not scrupled to call Leibniz a thief; and who therefore does not really deserve to be mentioned today in this hall."[27]

It would be wrong to think that, in their attempt to construct a body of rational relationships between theory and experience with matter as the fundamental assumption, Kelvin and Tait are so tied to experience in the question of the cognitivity of matter itself that they cannot grasp the subtlest forms of mathematical abstraction. While maintaining that deduction must come after the inductive phase and emphasizing the purely instrumental character of mathematics, Kelvin and Tait always keep in mind the advances in mathematics and in effect try to reckon with them. The two Scottish physicists are quite aware of the fact that there are unresolved enigmas and do not deny that before such enigmas one must say *ignoramus*. But they provide an optimistic justification for the *ignoramus* by arguing that nature will always be full of surprises and unexpected events. And the question of mathematics is never far from their minds, especially in connection with the concepts of space and time, which, after matter and force, are the fundamental concepts in their view of the world.

A good Newtonian must certainly have clear ideas about space and time. But it is also true that these clear ideas must be measured against the norm expressed by Tait in the sentence, "There is no such thing as absolute size." In Tait's opinion there are mathematical methods that are particularly suited to deal with the ideas of space and time. Such are the methods developed by W. R. Hamilton, and consist of algebra, understood as the "science of pure time," and geometry, designated as the "science of pure space."[28] Thus Hamilton's sciences can help us to forge a permanent bond between physics and a three-dimensional view of space that will conform to our perceptions. But the fact that this view of space has been accurately analyzed by B. Riemann and H. Helmholtz cannot be ignored. Their mathematical studies have introduced a fundamental problem without, however, finding a solution for it and, as Tait sees it, the problem is "whether space may or may not have precisely the same properties throughout the universe."

It is unnecessary to call attention to the fundamental nature of the problem embodied in Tait's sentence or to the fact that in the 1870s the relationship between physics and geometry was stated in just those terms. On the other hand, it may be instructive to see how Tait attempts to find a provisional answer to this question by once again relating its significance to the context of sensations and experience.

As Tait suggests, the new questions introduced by Riemann's abstract studies may be discussed by examining the different sensations felt by an imaginary being who lives in a two-dimensional world.[29] In this flat world, our imaginary being finds himself in general conditions that can be summarized by saying that he can live and move on either flat or curved surfaces. If this two-dimensional world has both flat and curved regions, our imaginary being, although constitutionally incapable of perceiving anything that may suggest to him the concept of a third dimension, "would certainly feel some difference of sensations in passing from portions of his space which were less, to other portions which were more, curved."

Is there an analogy between the sensorial differences felt by the two-dimensional body of the hypothetical being and the sensorial differences that could be felt by our bodies?

According to Tait, this possibility entails the abstract notion of a fourth dimension of physical space, but at the same time it implies something quite concrete in terms of experience. We travel through cosmic space along with the planets of the solar system. Nothing forbids a priori that this journey may bring us to regions of the universe "in which space has not precisely the same properties" we can currently ascribe to it. Thus, should we convince ourselves empirically that these new properties of space really exist, we could hardly deny that the new properties "will necessarily imply a fourth-dimension change of form in portions of matter."

In other words, the problem introduced by Helmholtz and Riemann leads Tait to envisage a situation in which a two-dimensional being moves about in a universe represented by a surface of varying curvature. And by analogy between that hypothetical universe and our real world we arrive again at the supremacy of empirical knowledge. The discussion of this problem undoubtedly forces the scientist into what Tait calls "mathematical reasoning of a transcendental character," but Tait has never denied that in drawing the necessary consequences of experimental principles theories may lead us "far from observation" and that certain highly abstract branches of analysis may therefore become useful. What Kelvin and Tait deny is that these abstract branches may be a form of a priori knowledge of the objective world. Even the most subtle and sophisticated inferences drawn from Riemann's geometry must submit to interpretations based on experience, and herein lies their potential value as instruments in the hands of the Newtonians.

The philosophical perspective from which Tait approaches Riemann's work takes no account of what Riemann considered characteristic of the learning process. In an essay published in 1867, "On the Hypotheses on Which Geometry Is Founded,"[30] Riemann reiterated the view he had already expressed in 1854 that investigations of a very general character were precious allies in the battle against "the too limited views" that hindered the advancement of scientific learning with "traditional prejudices" and claimed to find solid bases in the crude intuitions of ordinary life. In the fragments published posthumously in 1876 Riemann wrote that "our conception of nature little by little becomes ever more complete and precise, but at the same time further and further removed from superficial appearances."[31] Riemann too, then, thought that rational knowledge grows by successive approximations. But while he entrusted this growth to the tools of abstraction, and saw knowledge itself as ever more detached from surface appearances and from the teaching of the senses, Kelvin and Tait, on the other hand, sought an objective knowledge that even in its most general and abstract form would still be bound by the dictates of ordinary, immediate perception.

Thus the attempt of the two physicists to fight both the renunciatory doctrine of Emil Du Bois-Reymond and a theoretical physics they accuse of betraying empirical knowledge appears to be a well-thought-out project. Its goal is to save the natural sciences not only from the doctrines that would set for them insuperable limits, but from the theorists who tend to emphasize the virtues of mathematical knowledge over the certainties of experience.

While their opposition to renunciatory philosophies is based on a recognition of the element of irrationality they contain, it is far more difficult to identify with any precision Tait's (and Kelvin's) motives in fighting Boltzmann's physics. If the authors of the *Treatise* somehow manage to bring scientists like Fourier and Riemann back to the safe grounds of empirical knowledge, one might well ask why an analogous operation does not seem possible to them in the case of scientists like Boltzmann.

Thus if we wish to understand why certain scientists are accused of using mathematics as a substitute for thought, we have to delve a little deeper into the physical and philosophical conception of the world and of the relationship between science and nature that is embodied in the work of Kelvin and Tait. We already have some of the essential elements of this conception. The stucture of matter is not based on primordial elements to be discovered once and for all; on the contrary, matter is the inexhaustible premise to all empirical investigation, which advances by successive approximations without ever encountering any limit. Space and time are not concepts established once and for all but problematic notions that have to be further analyzed with Riemann's geometry, Hamilton's algebra, and sense experience. Prime causes are not the object of scientific inquiry, for the same reasons that justify the claim that matter has no ultimate and absolute structures. The only stable reference points in the investigation of the world are the principles of mechanics. Their stability, however, does not stem from a presumed ability of mechanics to passively reflect the nonexistent, ultimate elements of nature, but from the authority that Baconian rules lend those same principles.

Thus, before proceeding, it should be clear that the journey through the work of Kelvin and Tait is not a journey through the lands of a mythical continent invented by philosophers and christened with the generic name of mechanism. The itinerary we are forced to follow leads us elsewhere, and its stages do not correspond to those marked on fictitious maps by the historians who have tried to confine nineteenth-century physics within the philosophical walls of mechanism. Traditional philosophy has handed us an erroneous cartography: should we follow its directions, we would soon be lost in mythological wanderings.

The Destruction of the Clock-Universe

We will now proceed to take a closer look at Kelvin's and Tait's conception of the world and of the relationship between science and nature. Owing to the eclectic character of Scottish Newtonianism, their conception is neither unified nor homogeneous; we are confronted with a Newtonianism that has the structure of an archipelago rather than that of a continent.

In the name of Newton, in the 1860s Tait and W. J. Steele write a textbook on dynamics in which they speak of dynamics as the crowning glory of Newtonian thought.[32] A few years later, however, Tait writes in the magazine *Nature* that the concept of action at a distance is a useless blunder. And if that is what action at a distance is, then the molecular sphere of action is a delusion and the kinetic theories of gases, which are based on this very concept, are hopelessly false. *"The molecular sphere of action,"* Tait says, "does not convey any clear idea to the nonmathematical mind," and "the idea it conveys to the mathematical mind is instead hopelessly false." "In effect, I have no faith in attractions and repulsions that act at a distance between centers of force and according to different laws."[33]

The condemnation of one of the fundamental concepts of Laplace's Newtonianism is quite clear. Equally clear is Tait's stand in favor of the revolution started in England by Faraday: action at a distance must be replaced with action by contact. However, one has to make a distinction between Faraday's ideas and Maxwell's. The latter lead to a highly mathematical theory of the electromagnetic field that Kelvin will criticize for years, referring to it as the "so-called" electromagnetic theory of light.[34]

To Kelvin's methodology, for which Tait is often the spokesman, Maxwell's theoretical physics does not appear too different from that betrayal of empirical research that is best exemplified by Boltzmann's theories. With great skill and tenacity, Kelvin will therefore attempt to take apart "the Boltzmann–Maxwell doctrine," and will endeavor to prove it groundless by means of thought experiments[35] aimed at disproving the improper generalization that, in his opinion, Boltzmann and Maxwell have constructed. More on this point later. For the time being, let us just keep in mind that Tait and Kelvin hold to a Newtonianism radically different from that of a Laplace, or a Poisson, or an Ampère, in the sense that Tait and Kelvin develop on their own a physics of the continuum based on action by contact.

But what are the specific foundations of this physics? Obviously, it cannot be built solely on the motto that experience is the only source of knowledge. Much more is needed, namely, a detailed conception of the physical world, a system of noncontradictory statements capable of explaining phenomena by reducing them to manifestations of the principle of action by contact. All this cannot be achieved either with Bacon's philosophy and the glorification of true Newtonianism, or with a generic appeal to crucial experiments presumed to be unfailingly competent to eliminate erroneous theories.

If we are to start, as Tait does, from action by contact, then the universe must be a continuum. And if the universe is a continuum, the propagation of events in it must obey certain laws. Once discovered by observation, these laws must be amenable to mathematical formulation so that we can study their consequences and compare the latter to laboratory data. The rules of induction guarantee that the procedure is correct but never tell us which specific form it should take. They never tell us, for instance, the particular form in which we are to write the differential equations that will describe concisely the experience of a universal continuum.

There is an additional difficulty. The universe of action at a distance is a cosmic clock. Since every event is repeated in cycles, there is no history in such a universe. Everything comes back to itself, every natural process is reversible. But for Kelvin and Tait the universe is not a cosmic clock. Nothing repeats in the world because the world evolves in time. There are no cycles, because all phenomena are irreversible.

The fundamental empirical law of the universe does not rest on the time symmetry that obtains in abstract dynamics, but on the asymmetry that without exception prevails in nature. Experimental science, Kelvin says, has triumphed over the illusions of the mathematics of action at a distance between centers of force by showing that objects and organisms develop gradually according to a new law: the principle of degradation of energy. This principle, enunciated by Kelvin in 1851, almost forty years before the Tait–Boltzmann controversy, constitutes one of the scientific roots of Tait's accusations against Boltzmann.

It was in March 1851 that Kelvin laid the foundations for modern thermodynamics in the *Transactions of the Royal Society of Edinburgh*.[36] H. Davy's crucial experiment, Kelvin writes, has proved that caloric does not exist and that thermal phenomena must be explained through a dynamic theory of heat. Joule's experiments are "a perfect confirmation" of Davy's theses and, together with a radical reinterpretation of Sadi Carnot's ideas, they open the way to a new theory.

A very serious problem arose, however. The new theory was based on two principles of experimental origin, in the sense that they ultimately derived from observations on attrition in fluids, on the heat produced by electric currents, and on the operation of thermal machines. Although the empirical origin of the principles was very reassuring to Kelvin (who regarded their empirical origin as a proven and unquestionable fact), the problem of the relationship between the first and second principles was far less reassuring.

Whereas the first principle postulated the conservation of energy, the second implied a lack of symmetry in the evolution of natural phenomena. Only a few months earlier, Kelvin had judged this situation to be extremely worrisome.[37] He saw a contradiction between Joule's theses on energy conservation and Carnot's theories on the direction of phenomena, and reached the conclusion that this anomaly could be eliminated by a new formulation of Carnot's ideas that would not include detailed conjectures about molecular motions. Hence his critique of Rankine's and Clausius' attempts to deduce the two principles from hypotheses concerning molecular motions, and his confidence in a thermodynamics capable of describing phenomena without making too strong a commitment as to the underlying structure of matter. By proceeding in this manner, Kelvin could still consider himself a good Newtonian. On the basis of these "new foundations" Kelvin derived what he himself defined as the "remarkable consequences which follow from Carnot's proposition":

1. There is at present in the material world a universal tendency to the dissipation of mechanical energy. 2. Any *restoration* of mechanical energy, without more than an equivalent of dissipation, is impossible in inanimate material processes, and is probably never effected by means of organized matter, either endowed with vegetable life or subjected to the will of an animated creature. 3. Within a finite period of time past, the earth must have been, and within a finite period of time to come the earth must again be, unfit for the habitation of man, as at present constituted, unless operations have been, or are to be performed, which are impossible under the laws to which the known operations going on at present in the material world are subject.[38]

Kelvin thus sanctioned the end of a view of the world that perceived the universe as a cyclic machine. At the same time, the construction of scientific knowledge—and in his epistemological optimism Kelvin could only regard the growth of knowledge, if not as a construction, at least as a progressive stratification[39]—brought to light a troublesome dilemma: while the orthodox Newtonian method invoked mathematical methods in which the formal solutions describing motion generally did not change their form when the sign of the time parameter changed, the "new foundations" of thermodynamics were based instead on a physical principle of irreversibility. How could a theory of motion that did not distinguish past from future[40] be reconciled, within a unified framework, with a dynamic theory of heat that made a clear distinction between the two? And what problems would arise for the student of nature who realized that the extension of knowledge from laws of motion inductively certain led to other laws that were also inductively certain but implied irreversibility?

Kelvin firmly believed that the growth of the physical sciences was an endless succession of approximations that started from Newtonian principles. But it was equally clear to him that this growth had brought about the end of action at a distance and of universal irreversibility. It was still a growth by stratification, but to avoid paradoxes the role played in this process by Laplacian-type arguments had to be carefully re-examined.

For Kelvin this was the beginning of a long and complex intellectual journey. A journey that would end in 1896 with the harsh self-criticism Kelvin expressed at the celebration of his fiftieth year of teaching at the University of Glasgow, in response to the greetings coming to him from all over the world: "One word characterizes the most tenacious efforts I have persistently made for fifty five years in order to advance scientific knowledge: and that word is FAILURE."[41]

That journey encompassed the whole of physics of the second half of the century, engendering battles that affected geology, astrophysics, and biology as well. The Tait–Boltzmann controversy cannot be understood unless it is reconstructed in relation to this rich scientific background and, in particular, to Kelvin's dilemma on the relationship between the two principles of thermodynamics and the mathematical theories of motion, for which the solutions are invariant with respect to time reversal. Otherwise, this particular controversy might get lost in the fog of the *querelle* between old philosophies.

Science and Hypotheses

If Newton's equations of motion are symmetrical with respect to time, it is impossible to use them in discussing a system of many molecules—a gas —and deduce that the system will be irreversibly transformed. Whoever attempts to explain the two principles of thermodynamics with mechanical models of molecular motion embarks on a hopeless venture. And should he succeed in expressing irreversibility starting from such models, then he must have made a mistake in his deductions, or else he must have introduced an ad hoc principle in the mathematical theory.

Not only are those models based on the notion of action at a distance, by now demolished; not only do they try to explain the properties of macroscopic bodies by attributing the same properties to hypothetical molecular or atomic entities, thereby bringing back into vogue "Lucretius's monstrous hypothesis," and transferring to the level of microphysical entities the problems unresolved at the level of the objects observable by the senses or the microscope;[42] worst of all, those models induce scientists to hide the impotence of their assumptions behind an ever more abstract mathematics. Consequently, the proponents of the kinetic theories of matter are dogmatists who hinder the progress of physics, obscure the clarity of mathematics, and becloud the sky of dynamics.

This is Kelvin's position with regard to the theories of Clausius, Maxwell, and Boltzmann. In the 1860s, while Clausius was working out a mathematical formulation of the second principle of thermodynamics and a physical basis for the famous thesis that "the entropy of the universe tends to a maximum,"[43] Kelvin answered by attacking on many fronts. In the first place, the statements of the kinetic theory of matter rested in his opinion on unacceptable hypotheses and on models that could not be experimentally verified. In the second, the thesis of entropy was nothing more than a useless and harmful rehashing of the principle of dissipation of mechanical energy. In the third place, it was improper to expand the scope of the second principle of thermodynamics to the point of having it predict the thermal death of the whole universe. The following is an excerpt from an article written by Kelvin for *Macmillan's Magazine:*

The second great law of Thermodynamics involves a certain principle of *irreversible action in nature*. It is thus shown that, although mechanical energy is *indestructible*, there is a universal tendency to its dissipation, which produces gradual augmentation and diffusion of heat, cessation of motion, and exhaustion of potential energy through the material universe. The result would inevitably be a state of universal rest and death, if the universe were finite and left to obey existing laws. But it is impossible to conceive a limit to the extent of matter in the universe; and therefore science points rather to an endless progress . . . than to a single finite mechanism, running down like a clock, and stopping forever.[44]

Thus the infinity of matter and the endlessness of history characterize the complex dynamics of an objective world that is no longer conceivable as the works of a clock. And if the objective world has a history without end, one can no longer speak of its future thermal death. The conjectures about an entropy tending to a maximum value are just so much meaningless prattle.

Kelvin's arguments, while resting on detailed calculations, had a very broad import. In a general way, they not only challenged the credibility of the investigations into the structure of matter carried out in the context of kinetic theories, but struck hard at all those research programs in geology and biology that needed to assume near-infinite geophysical times in order to explain the evolution of organic or inorganic matter. One of the victims of Kelvin's ax was, for instance, Darwin's theory. As Kelvin saw it, that theory was remarkable for its "extraordinary futility." The meaning of Tait's accusation against Boltzmann will become yet clearer if we recall the recurring theme in Kelvin's criticism of the author of *The Origin of Species by Means of Natural Selection,* namely, Kelvin's contention that Darwin made use of improper hypotheses, thus betraying the spirit of true Newtonian research. As Kelvin stated in his presidential address at the 1871 meeting of the British Association, the true naturalist knows that *"the essence of science . . . consists of inferring antecedent conditions, and anticipating future evolutions, from phenomena which have actually come under observation."*[45]

It is this appeal to experience, now as ever, that underlies his criticism of those who fail to base knowledge on phenomena that have actually come under observation. If we read Kelvin in Kelvin's writings rather than in the pages of Duhem (or of any other historians and philosophers of science devoted to mechanism), we will find that he seeks to extend knowledge, not on the basis of models, but, on the contrary, on the basis of what he considers a phenomenon actually observed. We will also find that he fights hard to eliminate all those theories that betray Newtonianism by assuming as premises conjectures about what is unobservable. The kinetic theory of matter and Darwin's evolutionary theory are in his opinion two cases in point.

But is it always possible to build theories from observed phenomena? In Kelvin's youthful writings the answer is in the affirmative. A thoughtful student of Fourier's work, our author defends his mathematics and tests its power by extensive use not so much of models as of "analogies." And in this we find, above all, an echo of the views on physics expressed by G. Lamé, whose mathematical work has left a deep impression on the young Kelvin. In a treatise on physics,[46] Lamé had written that the natural sciences must be freed of "vague and by now sterile hypotheses" and of "uncertain and metaphysical theories." Scientific teaching must emphasize "the practical scope and the philosophical character" of physics and of the other sciences. It is to be hoped that "one day it will be possible to have the teaching of physics consist only of an exposition of the experimental and observational procedures which lead to the laws of natural phenomena, without it being first necessary to formulate premature, and often harmful, hypotheses concerning the ultimate causes of these natural phenomena. It is precisely to this positive and rational state that science must be brought." According to Lamé, analogies are the foundations of scientific research: the method of analogies is a very effective tool "for deriving more or less probable inferences as to the identity of the causes" and must accompany observation and the art of experimentation to make possible efficient research into the phenomena.

In Lamé's treatise physics is presented as a "critical research" that always starts from experimental and observational data and advances by use of the analogical method understood as a form of induction. In Lamé's view also, mathematics comes after. "When the laws discovered empirically can be translated into numbers, we apply calculus to them, and mathematical analysis provides all the consequences that can be derived from these laws, which are presumed true." Thus, it is fair to recognize the merits of an abstract thought that operates within norms, and to consider mathematical analysis as that "which alone can make reasoning infallible." According to Lamé, however, the rational process is divided into two distinct branches, which are based on different criteria: experience on one side, and deduction on the other. The former gives us a partial but sure knowledge of what exists in the world; the latter constitutes a precise language into which the scientist translates what he has previously learned.

It goes without saying that Lamé admires Fourier's theories. Fourier, too, had celebrated the virtues of mathematics as an all-powerful language. But while Fourier saw in mathematics a reflection of the structure of the world, Lamé regards mathematics as a language that describes the structure of the world as discovered by induction. Once again, Fourier has been reinterpreted in such a way as to equate the constant and simple laws, discovered by observation, with the general laws deduced by analysis.

Lamé's views thus appear to have had a significant impact on Kelvin's scientific formation. Kelvin mentions Lamé's memoirs in the papers of his early twenties, when he is studying Fourier, Chasles, and Gauss.[47] In these writings Kelvin has already departed from the model making of the French school that had culminated in Laplace's celestial mechanics and in Poisson's elegant models. Lamé himself gives voice to the irremediable split over the question of models that has caused the decline of the French school of mathematical physics and has found in Fourier a tireless advocate. In the *Leçons sur la théorie analytique de la chaleur*, Lamé passes judgment on models with these words: "We have to admit it: educated at the school of Laplace, neither Poisson nor Cauchy was capable of thinking that it might be possible to formulate a mathematical physical theory without assuming any laws at all."[48] Implied in this sentence is the contrast between Poisson's mathematical physics and Fourier's. Assuredly, a contrast exists and revolves around the role—positive, negative, or irrelevant—that is allotted to models. However, Lamé does not do Poisson full justice on the question of mathematics' role and its importance. In actuality, when Poisson discusses the application of mathematics to physics he makes statements that are nearly the same as those considered correct by Lamé, in the sense that Poisson is also convinced that mathematics should "come after."

In the *Théorie mathématique de la chaleur,* published in 1835, Poisson writes that a physicist should first formulate a hypothesis concerning the nature of thermal radiation such as can be inferred from observation, and that only later should mathematical reasoning be applied to that hypothesis. During this process, mathematics neither adds nor subtracts anything from what is already known. Also for Poisson, then, to formulate a physical theory in mathematical terms means to translate a body of given knowledge into symbolic language.[49]

Thus, the complexity of the interpretations given to the question of models in physics appears to be the result of the even more complex and unresolved question of the cognitive function that can be rationally assigned to mathematics, in other words, whether mathematics is a language and an instrument to be used only after a discovery has been made, or whether it is also capable of discovering facts on its own and of playing more than the merely instrumental role allotted to it by those who believe in the primacy of experience.

During the period spanning the work of Poisson and Lamé, mathematics applied to the empirical sciences finds itself in a dramatic situation. We need only recall the violent disputes that arise in England between logicians and mathematicians during the fourth decade of the century. While young Kelvin is studying at Glasgow and Cambridge, William Whewell, John Herschel, and George Peacock strive to introduce into England the refined mathematics of the continental school. Their attempts are tenaciously resisted by logicians and philosophers like William Hamilton. It is Hamilton who writes in the *Edinburgh Review* of January 1836, "An extensive study of the mathematical science . . . absolutely *incapacitates the mind,*" and again, "The mathematician knows nothing."[50]

The following excerpt from Hamilton's article is quite interesting in this respect: "Mathematicians are also infested with an overweening presumption or incurable arrogance, for, believing themselves in possession of demonstrative certainty in regard to the objects of their peculiar science, they persuade themselves that, in like manner, they possess a knowledge of many of the things beyond its sphere."[51]

It would be wrong to reduce the scope of Hamilton's violent polemic to an isolated case. The stiffening of the antimathematical posture is an institutional problem for English science in those decades, and is part of the scientific and philosophical travail that will result in the formation of the mathematical school of Cambridge and in the establishment of new directions of research in English physics. A fundamental role in this context is played by the scientific and philosophical mediation carried out over the years by J. Herschel and, in particular, by a very successful book that he published in 1830 and that had a great influence on physicists like Kelvin, Tait, and Maxwell. We will come back to this shortly. Right now let us continue with our examination of the problems that arise in physics when one sets the condition that theories must start from observed phenomena rather than from a priori models. This road will take us, on the one hand, to a new insight into Tait's stand against Boltzmann, and, on the other, to Herschel's arguments.

Analogies and Common Sense: The Virtues of Induction

As we have just seen, young Kelvin firmly believes that analogies are the sovereign virtues of the inductive method and the guarantees of a knowledge that, even when expressed in mathematical form, will thus have the support of empirical evidence. But his faith in analogies does not stem from mechanism; rather, it is based on the ideas of those physicists who are opposed to Laplace, that is to say, on the arguments that part of the French school brings to bear against the use of mechanical models, against pretensions aimed at raising dynamic theories above experience, against attempts at reinstating the supremacy of metaphysical hypotheses in the natural sciences, and against the search for prime causes by mathematical physics. It is also based on Kelvin's growing awareness of the evolutionary nature of the universe. The young physicist, born in Belfast in 1824, is keenly interested in the debates on the history of nature which pervade European culture and whose multiple roots reach into the fields of biology, geology, and cosmology. Finally, as we noted earlier, it must be remembered that all these factors operate at a time when European physics is trying to find rigorous norms that will enable it to choose between action at a distance and action by contact. In his early writings Kelvin has already made his choice. This choice is not based on unfounded opinions about physics or on sudden intuitions but on a thoughtful study of Faraday's work and of the view of the physical world that is embodied in it and that leads to a conception of the universe as a continuum crossed by lines of force of varying curvature.

From this complex and intricate background gradually emerges in Kelvin's work the dominant idea that the objective world is a plenum of infinite dimensions. Wherever normal matter is so rarified as to be no longer observable with the senses or with laboratory devices, there still exists a "something," an infinitely extended entity that fills all the pores and cavities of bodies. The history of phenomena and the motion of the stars takes place in this medium—the ether. If the world is a continuous structure wherein actions propagate by contact, it stands to reason that there must be something that allows them to propagate even where matter cannot be observed.

What is the proper approach to the investigation of this universal entity? In 1847 Kelvin states in one of his most brilliant scientific papers that the key to this research lies in Faraday's discoveries in electromagnetic induction, which suggest "the idea that there may be a problem in the theory of elastic solids corresponding to every problem connected with the distribution of electricity in conductors, or with the forces of attraction and repulsion exercised by electrified bodies."[52]

This is a peculiar variation of the analogical method. In his previous papers Kelvin had discussed some aspects of the theories of electricity and of thermal phenomena starting from the discovery of analogies between the mathematical tools that could be used in both theories. He now generalizes that method to the point of connecting by analogy Faraday's empirical problems to the sophisticated mathematical problems of the theory of elastic solids formulated by G. Stokes. Faraday, on the other hand, is opposed to Kelvin's procedure both because it reintroduces the very concept of ether he wants to "dismiss" from physics[53] and because he cannot grasp the implications of the abstract treatment developed by Kelvin.

The 1847 essay ends abruptly, however, and in the subsequent decades Kelvin follows different research paths. The reasons for this diversification are many. In the first place, we must remember that in 1847 Kelvin is not yet convinced of the validity of Ampère's hypothesis concerning the electric nature of magnetism. Not until 1856 will that hypothesis be incorporated into Kelvin's physics, and even then it is placed in an entirely different mathematical context and a view of physics that is dominated by Kelvin's rejection of "many . . . statical preconceptions regarding the ultimate cause of apparently statical phenomena."[54]

His world is no longer continuous and static by now, but infinitely continuous and moved by an inner dynamics. Accordingly, he is no longer concerned with the mathematics of an elastic solid representing the ether, but with a more complex formal treatment of hydrodynamics.

It is from these new premises that Kelvin speaks of Lucretius' "monstrous hypothesis" and works for almost thirty years at the development of a new concept of the structure of matter: the vortex-atom.

The development of this concept implies a very complex network of relationships between statements on phenomena and mathematical analogies. Moreover, it is not a sudden development but a drawn-out process spanning three decades. Some of its aspects recall Rankine's attempts, after 1848, to deduce thermodynamics from hypotheses of a vortical structure of matter. In the gradual definition of the vortex-atom there is a confluence of ideas taken from interpretations of the second law of thermodynamics, from certain formal aspects of the theory of electrodynamics, from Helmholtz's mathematical work on fluids subject to vortical motions, from considerations on the application of Helmholtz's theory to meteorology, and from certain phases of Riemann's studies on geometry.[55]

The initial results of this process of formulation of the concept of vortex-atom appear quite positive to Kelvin and Tait. The latter provides Kelvin's concept with some sort of empirical basis by means of simple experiments that show the surprising properties exhibited by smoke rings in their motions and mutual interactions.[56] As a result, there is a growing belief that the vortex-atom is the "real" atom, and that its properties will permit the elimination from the world of physics both of Lucretius' hypothesis and of what Kelvin calls the rashly worded statements of some chemists about the molecular structure of matter.

Matter now appears not as a void wherein Boltzmann's or Maxwell's atoms casually meet, but as a plenum: an immense, perfect fluid within which annular structures, whose motions are governed by Helmholtz's complex differential equations, revolve and collide.

In 1867 Kelvin writes, "The only pretext seeming to justify the monstrous assumption of infinitely strong and infinitely rigid pieces of matter, the existence of which is asserted as a probable hypothesis by some of the greatest modern chemists in their rashly-worded introductory statements, is that urged by Lucretius and adopted by Newton—that it seems necessary to account for the unalterable distinguishing qualities of different kinds of matter."[57]

But Helmholtz's rings and vortices do imply "an absolutely unalterable quality in the motion of any portion of a perfect fluid in which the peculiar motion which he calls '*Wirbelbewegung*' has been once created." Hence *Wirbelbewegung* is the unalterable and stable vortical motion that can replace the atom envisaged by Clausius and Maxwell. This substitution is not only possible but necessary, because the number of assumptions needed to formulate the new theory is much smaller than the number required for the usual kinetic theories. The new theory only needs to invoke inertia and the incompressibility of occupied space. This is the premise to the task of formulating a purely mathematical analysis of the "mutual action between two vortex rings of any given magnitude and velocities." This analysis, Kelvin remarks, entails "a perfectly solvable mathematical problem" and poses new questions "of an exciting character."

As Kelvin himself says, we are now confronted with "an intensely interesting problem of pure mathematics." The fact that it is solvable does not mean that it is easy. On the contrary, "The analytical difficulties which it presents," even in the case of a simple Helmholtz ring, "are of a formidable character, but certainly far from insuperable in the present state of mathematical science."[58]

At this point it is essential to put two types of questions into the right perspective. The first concerns Kelvin's mathematics; the second concerns the notion of mechanical model in connection with, on the one hand, the kinetic theory of gases, and, on the other, the explanation of the structure of matter in terms of vortex-atoms.

Kelvin's praise of Helmholtz's mathematics and of the explicative power of hydrodynamic equations, coupled with his references to Riemann's geometrical studies, might suggest that our physicist has modified his ideas on the subordinate role of abstract language to the extent of espousing a view of mathematics similar, for instance, to that held by Riemann. But if this were the case, considering that the quotations we referred to earlier are taken from a paper of 1884, it would be very hard to understand why a short time later Kelvin should induce Tait to undertake that particular campaign against Boltzmann's theories.

In actuality, the changes occurring in Kelvin's dictionary with the passage of time and the influx of new theories are changes that Kelvin himself attempts to reduce to commentaries on the norm that experience is the only source of knowledge. The need to conceive of mathematics as a language and an instrument is also subject to evolution. Kelvin may be pleased to describe philosophers as hunters of soap bubbles and Tait may ridicule Hegel's obscurity or Du Bois-Reymond's renunciation, but, in fact, they both manipulate the concept of mathematics by, on the one hand, taking careful notice of what the mathematicians are doing, and, on the other, by making it conform philosophically to their view of the primacy of experience. In this respect Tait's and Kelvin's dictionaries have strong ties with a certain current of British culture that celebrates common sense. Mathematics is an instrument, Kelvin writes in 1883, because it is closely connected to common sense:

> Do not imagine that mathematics is harsh and crabbed, and repulsive to common sense. It is merely the etherealisation of common sense. Consider . . . the beautiful and splendid power of mathematics for etherealising and illustrating common sense, and you need not be disheartened in your study of mathematics, but may rather be reinvigorated when you think of the power which the mathematicians, devoting their whole lives to the study of mathematics, have succeeded in giving to that marvellous science.[59]

We have remarked earlier on the connection Kelvin sees between analogies in general, mathematical analogies in particular, and the basic norms of induction, which, to Kelvin, is the supreme tool in the discovery of the laws of nature. We now see how the connection between mathematics and induction is made to include common sense.

The fact that the splendid edifice built by mathematicians deals with abstract concepts by no means implies that mathematics is useless to the empirical sciences. Applied mathematics redeems itself in the field of knowledge because it is tied to the inductive method inasmuch as it aids induction in obtaining those details that are deducible from given physical laws. And from this point of view it is also tied to the confirmations of common sense. However general, in a purely formal sense, a series of deductive inferences may be, it must always be amenable to immediate translation into the language of common sense. Hence mathematics is "the etherealisation of common sense."

In advancing this thesis, Kelvin presents a comprehensive view of his ideas on the theory of knowledge. The question he asks himself in 1833 is very explicit: "What are the means by which the human mind acquires knowledge of external matter?" The answer is unequivocal: knowledge is acquired through the senses and only the senses. This is the key to understanding Kelvin's view of the relationships between mathematics and common sense and between mathematics and induction. It should be noted that Kelvin himself points out one of the cultural sources from which he derives his view of mathematics as the idealization of common sense, namely, the common sense philosophy elaborated by Thomas Reid in the second half of the eighteenth century.

Thus we begin to see a network of relations that, starting from Tait's criticism of Boltzmann, reaches into the physics of Kelvin and Tait and branches out into the methodology of the French physicists opposed to Laplace's interpretations and into Herschel's reflections on experience. It is an itinerary that leads, through Herschel, to Thomas Reid. This itinerary passes through a few variants of empiricism but never crosses the imaginary provinces of mechanism.

This means that to understand Tait's polemic against Boltzmann we have to pursue a winding course through the levels of several interacting dictionaries rather than simply make an analysis of the method of mechanistic philosophy.

Kelvin and the Models

As we have seen, Boltzmann's English critics believe that a good Newtonian may formulate a physical theory in highly mathematical terms provided he does not go so far as to break the bonds that must exist between formal deduction and common sense. If these bonds are somehow broken, one falls into the fallacy of constructing a theory removed from the senses and founded on preconceived assumptions. Such a theory would be a mere mathematical exercise rather than a branch of empirical knowledge.

This point of view is certainly present in Tait's criticism of Boltzmann, yet does not suffice to fully understand that criticism. As a matter of fact, in the theories judged by Tait as mere symbolic games Boltzmann makes extensive use of models in which molecular physics is studied in connection with molecular dynamics. Since Kelvin and Tait are continually making models of a mechanical nature, Boltzmann should actually appear to them as an ally. The very concept of vortex-atom is closely related to a particular area of mechanics, in that the behavior of the vortex-atom is determined by Helmholtz's hydrodynamic equations.

Thus, if we were to stage the controversy between Boltzmann and the English physicists on these grounds, the result would inevitably be a great deal of perplexity. One might wonder whether the dispute between Tait and Boltzmann was, after all, nothing but an internal quarrel among mechanists, a clash between proponents of models that, although different, could still be unified in one mechanistic matrix.

We can give an affirmative answer to this question only if we are willing to argue that the mechanical explanation of a phenomenon is the same as an explanation by means of a mechanical model. If by mechanical explanation of a phenomenon we mean the construction of a mechanical model of the phenomenon in question, then there are no insuperable discrepancies between Tait's conception and Boltzmann's.

What we must analyze, therefore, is the following point: Is it or is it not true that according to Kelvin and Tait the understanding of a phenomenon is the mechanical explanation of it, and that the mechanical explanation is the construction of a model?

The identity of "understanding" and "making a mechanical model" is often judged to be at the heart of mechanistic thinking. In a celebrated statement made by Kelvin in 1884 while lecturing at the Johns Hopkins University in Baltimore,[60] our author seems to leave no doubt on the matter. First Kelvin says, "It seems to me that the test of 'Do we or do we not understand a particular subject in physics?' is 'Can we make a mechanical model of it?' " Then he adds, "I never satisfy myself until I can make a mechanical model of a thing. If I can make a mechanical model, I understand it. As long as I cannot make a mechanical model all the way through I cannot understand." Later on, however, Kelvin states that mechanical models by no means reflect reality: the phenomena we make models of are not the same things as what is discussed in the models. We can make models about the ether, but in discussing such models we must never forget that it is a discussion in the realm of the "as if." Models are not copies of reality but imitations. And, like all imitations, models are not unique. One can make different models of the same phenomenon, and the choice among them depends on several factors, such as simplicity, for instance, meant as economy in the number of necessary assumptions and judgment concerning the specific dynamics to be used.

For Kelvin, then, the logic of the relationship between understanding and models operates on two levels: on the one hand, there is the need to make mechanical models, and, on the other, the awareness of the fact that models are not copies of specific segments of reality.

Consequently, Kelvin's model is only meant to give us an idea of how a phenomenon would occur if it were structured in such and such a way. In sum, we are in the realm of the "as if" rather than in an imaginary mechanistic temple where models passively reflect nature. And the realm of the "as if" is for Kelvin the very realm where successive approximations are built. Hence the need for models, which for Kelvin are devices for rationally transferring certain mathematical solutions from given and well-founded theories, and for testing those solutions in theories that are being developed. We see this happen in Kelvin's work when the formal apparatus of hydrodynamics is applied to the problems of continuum physics. Kelvin's purpose is to use Helmholtz's solutions, already known, to explore the grounds of a new theory of electromagnetic phenomena that he is developing. As a result, Kelvin can now investigate the analogies between the dynamics of vortices and the phenomenology of magnets,[61] and engage in model making in the sense of engaging in research.

No doubt, it is a type of research that follows a particular approach in which scientific explanation is seen as explanation based on the mathematics of motion. But the use of this particular approach cannot be equated to that complex of conceits which is normally defined as mechanistic understanding in terms of models. It is one thing to discuss a group of phenomena by seeking precise guidelines in formal solutions that have already shown a certain validity in other theories, and quite another to draw a few interconnected gyroscopes on the blackboard and say that the ether behaves "as if" it were made of gyroscopes. There are differences both in form and in content. In 1928 Dirac started from the Hamiltonian of Klein and Gordon to arrive at the relativistic theory of the electron. Should we then say that Dirac's physics is mechanistic because it is based on a development of W. R. Hamilton's mechanics? Or should we instead conclude that Dirac was not a mechanist because he interpreted the Hamiltonian in a nonmechanistic manner? Are we not in effect reducing the problems of physics to questions of personal taste?[62]

Of course, there is another use of models, namely, for didactic purposes. Like most physicists of his time, Kelvin made quite an extensive use of models as an aid to understanding the physical meaning of specific formal solutions. The extraordinary complexity of the mathematics of the ether profited from the study of various mechanisms and devices, such as systems of levers, winches, springs, and gyroscopes. But only the naiveté of a poet would believe, and try to make us believe, that Kelvin actually saw a faithful copy of the real structure of the world in the devices he himself defined as very crude. Those devices were just simple, concrete representations of what Kelvin called physical ideas embedded in the intricate deductive steps of a particular theoretical structure; they were not images of the things discussed in that theory.

In other words, the representation of the ether which Kelvin obtains by drawing a number of gyroscopes on the blackboard is analogous to the representation we would obtain today by drawing a circle with a point in the middle and calling it "the hydrogen atom." It may be objected that there is a fundamental difference between the two representations: modern physicists *know* that the hydrogen atom is not made the way it is drawn on the blackboard, since current physics has rid itself of the mechanistic delusions that vitiated the representations of nineteenth-century physics.

Such a thesis, however, would imply that the greatest physicists of the nineteenth century were not competent to distinguish between didactic tools and the structure of the world. The idea is so absurd that Kelvin would have loved to use it as a proof of the fact that philosophers are often engaged in chasing soap bubbles.

Let us return now to the first use of models, which consists in the transfer of specific mathematical solutions from a given theory to one that is being developed. In the case of the vortex-atom the purpose of the transfer operation is to formulate an explanation of the structure of the continuum; this explanation is thus predicated on Helmholtz's equations and on the stated intent never to violate the necessary ties between mathematical abstraction and common sense. Before we go any further, we should repeat once more that in Kelvin's opinion these ties are essential if mathematics is to be the idealization of common sense and if the meaning of the theory is to be constantly submitted to the judgment of experience and the test of induction.

It is quite important to identify the reasons why the development of the theory of vortex-atoms—which was meant to replace the atomistic theories and to establish a new and general dynamic theory of the universe —ground to a halt in the last decade of the nineteenth century. The fact that in 1896 Kelvin admitted the failure of his research program and accepted atomistic physics[63] marked a turning point of the greatest significance not only in the thinking of the septuagenarian "second

Newton," but in the history of the physical sciences. The causes of this change of direction are therefore quite relevant.

The causes of Kelvin's failure, and his awareness of it, have their origin in mathematics. The theory of the vortex-atom rested on an essential condition, namely, a condition of stability of the vortical structures animating the universal continuum. In the light of Helmholtz's mathematics, the continuum—a perfect fluid—was endowed with what Kelvin had defined in 1867 as an "infinitely perennial specific quality." "To generate or to destroy *'Wirbelbewegung'* in a perfect fluid can only be an act of creative power."[64] But this stability did not depend on specific ad hoc hypotheses: it was a mathematical consequence of Helmholtz's theory.

The fundamental question of stability was thus tied to Helmholtz's theorems. During the second semester of the 1891–1892 academic year, H. Poincaré gave a series of lectures on Helmholtz's theorems at the Faculty of Science in Paris. These lectures, published in 1893 under the title *Théorie des tourbillons*,[65] involved a re-examination of the theory of vortices that led to a generalization of Helmholtz's theorems and to a distinction between permanent motion and stable motion. This accurate analysis of the conditions of stability had a devastating effect on Kelvin's physics: in effect, there were no sufficient guarantees for the "infinitely perennial specific quality" that was the foundation of the vortex-atom.

Poincaré was aware of the import of such a conclusion. In the brief introduction to the *Théorie* of 1893 we read that the equations of vortical motions doubtless exhibit "a certain formal analogy to the equations of electrodynamics." Such an analogy "naturally leads to the establishment of parallels between the two theories, and in some cases has permitted the deduction of the solution of a problem that has arisen within one theory starting from a problem solved within the other." Helmholtz's fundamental theorem, however, "is fully applicable only to the motion of fluids in which there is no attrition, and which have a temperature either uniform or dependent only on pressure." When conditions differ even in small measure, the theorem is no longer applicable except as a first approximation.

The introduction to the *Théorie* is very explicit with regard to Kelvin's theory: "One has even attempted to find the mechanical explanation of the universe in the existence of these vortical motions. Instead of representing space as occupied by atoms separated by immense distances as compared to their dimensions, Sir William Thomson [Kelvin] holds that matter is continuous, but that some portions of it are animated by vortical motions which, as a consequence of Helmholtz's theorem, must retain their individuality."

The failure of Kelvin's research is embodied in these sentences. By generalizing Helmholtz's theorem, Poincaré has identified the limits inherent in the physics of vortices, and the latter, through its most prominent spokesman, admits to having reached a dead end. Its failure, in sum, is due to internal reasons for which a rational justification can be found, and in 1896 Kelvin makes his choice in accordance with rigorous norms. It is interesting to note that these norms are not weakened by the temptations of "conventionalism." It is not possible for Kelvin to amend the theory of the vortex-atom and save it from mathematical refutation, in the sense that he does not have a free choice among the concepts or postulates that might ward off Poincaré's attack. To rescue a theory it is not enough to make small readjustments that will restructure it and place it out of range. To do this, the concepts or postulates or laws to be freely chosen according to conventionalistic moves would have to be already formulated and available in the store of knowledge. And herein lies the unresolved problem in the strategy of every conventionalism: in order to have a free choice among the various propositions with which to attempt the rescue of a theory at all costs, these propositions must already be established. Though at times the choice appears free, this is not the case for the theoretical work that leads to the determination of what can or must be chosen. The choice between different notions is often restricted only by weak conditions, but the construction of the theory is planned and executed according to norms historically given and rationally expressed.

Thus, when the weapons of mathematics attack a physical theory, there is no methodological freedom so ample as to permit the rescue of the theory under attack by means of a topical treatment of its superficial aspects. If the attack is directed at the heart of the theory, the outcome of the battle cannot be decided at the level of methodological decisions, because these decisions do not affect the core of the theory itself. In the case of the vortex-atom there is no rational line of defense. Kelvin's failure does not stem from a methodological fiat, just as the identification of what was the core of the theory itself did not originate from a methodological fiat. Both decisions were taken on mathematical grounds.

In the early 1890s Poincaré's arguments cannot be refuted precisely because they do not leave room for free moves. Kelvin's conception of mathematics as the idealization of common sense by no means implies that a mathematical proposition amenable to empirical confirmation can be saved when it meets with unanswerable objections in the realm of deduction.

Reinterpretations: W. R. Hamilton according to Tait

If we ignore the metaphors about mechanism, the history of the vortex-atom clearly reveals the gap perceived by Boltzmann's critics between experience and deduction. It is a gap that separates the act of induction from the act of deduction, in the sense that the two acts are believed to be governed by different rules. In order to understand the meaning of the criticism leveled at Boltzmann, we must now analyze this dichotomy in the physics of Kelvin and Tait.

In their physics there are some fixed points that we discussed in the preceding paragraphs and that we shall now summarize as a set of theses.

The first thesis is that there are no absolute, ultimate elements in nature, and consequently no absolute physical theories. Instead of absolutely true theories—mechanics—there exist sure principles: the law of gravitation, Hooke's law, and so on.

The second thesis deals with the certainty of those principles. They are certain because they are the products of generalizations by induction, that is, the products of cognitive procedures guaranteed by experience as the only source of knowledge rather than by their conformity to mechanical models reflecting the structure of the world.

The third thesis is that knowledge is an extension by successive approximations, where mathematics is a tool for deriving logically necessary consequences from empirical principles already discovered by induction.

From the first three theses one obtains the fourth, which states that the growth of our physical knowledge has no limits either in hypothetical ultimate elements (no longer analyzable) or in history. There is scientific progress and throughout its course mathematics is the servant of induction. As a tool, mathematics has certain constraints, and as an idealization of common sense it is closely tied to nature. As Kelvin and Tait write in the *Treatise*, mathematics applied to physics is on firm ground when it enables us to elaborate methods that work "in a practically sufficient manner" and in which "Nature's own solutions" come into play.

These four theses lead physicists to view the scientific ideas of matter, space, and time not as absolute concepts but, as Tait remarks, as problematic concepts that are susceptible to possible modifications of sense experience itself, for instance, in the event that our planet should enter non-Euclidean regions of the universe.

Clearly, the Newtonianism of the *Treatise* does not have fundamental elements in common with the mechanistic conception of the world. The one absolute component in the rational investigation of the world is the inductive method, and this method consists not of a list of precise norms but of an appeal to turn directly to nature without prejudices and without presuming to deduce the natural order a priori. As Kelvin writes, "The essence of science . . . consists in inferring antecedent conditions, and anticipating future evolutions, from phenomena which have actually come under observation." If, in drawing inferences and making predictions, our mathematical tools reveal anomalies in the process of deduction, then there is no doubt that the deductive part has to be revised without any ad hoc modifications. Crucial experiments play a determining role in the choice between different theories and conjectures; but in the theoretical field, since there is a dichotomy between theory and practice, crucial mathematical objections, that is, the discovery of errors in reasoning, are sufficient to invalidate the theory.

The conception of mathematics that emerges from this view of the world and of the relationship between science and nature is quite evident in some of Tait's mathematical works. Particularly interesting in this respect is *An Elementary Treatise on Quaternions,* published in 1867, in which Tait discusses some of the general concepts of Hamilton's mathematical physics.[66] W. R. Hamilton's mathematical physics rested, on the one hand, on an explicit denunciation of the irrelevance of models in the cognitive process, and, on the other, on the need to study natural phenomena by means of highly abstract notions. Thus, there were similarities between Hamilton's science and Fourier's program. Hamilton's criticism of models is well known. We will only recall that in founding mathematical optics Hamilton asserted that the new field had its own validity as a "separate study" fully independent of any Newtonian conjecture or any set of hypotheses, such as Huygen's, concerning the nature of light.[67] Hamilton's reflections on algebra and on the extent of its applicability to the physical sciences, although perhaps not so well known, are even more explicit about pure and applied mathematics. In his fundamental work on the theory of conjugate functions,[68] published in 1837 but already partly formulated in 1833, Hamilton developed the theme of algebra as the science of pure time—a theme that would later be elaborated by Tait.

How is this science to be understood? According to Hamilton, the meaning of the study of algebra depends on the viewpoint of the individual scientist. In essence, there are three views of algebra, which he defines as the "practical," the "philological," and the "theoretical." In the practical view algebra is considered an instrument, or a system of rules, and is judged only from the standpoint of its usefulness. The philological view regards algebra as a language and analyzes its imperfections as though the latter pointed to anomalies in the symbolic notation or in the "symmetrical structure of its Syntax." The theoretical viewpoint, on the other hand, has a specific purpose: "The thing aimed at, is to improve the *Science*, and not the Art nor the Language of Algebra. The imperfections sought to be removed, are confusions of thought, and obscurities or errors of reasoning; not difficulties of application of an instrument, nor failures of symmetry in expression."

As Hamilton remarks, there are certainly very serious imperfections and difficulties, for instance, with regard to imaginary numbers. But it would be very hard to construct a science on such difficulties, "though the forms of logic may build up from them a symmetrical system of expressions, and a practical art may be learned of rightly applying useful rules which seem to depend upon them."

Hamilton is fully aware that some scientists are opposed to the theoretical view of algebra:

So useful are those rules, so symmetrical those expressions, and yet so unsatisfactory those principles from which they are supposed to be derived, that a growing tendency may be perceived to the rejection of that view which regarded algebra as a *Science, in some sense analogous to Geometry,* and to the adoption of one or the other of those two different views, which regard Algebra as an *Art,* or as a *Language:* as a System of Rules, or else as a System of Expressions, but not as a System of *Truths,* or results having any other validity than what they may derive from their practical usefulness, or their logical or philological coherence. Opinions thus are tending to substitute for the Theoretical question,—"Is a Theorem of Algebra *true?*" the Practical question,—"Can it be *applied as an Instrument,* to do or to discover something else, in some research which is not

Algebraical?'' or else the Philological question, — *"Do its expressions harmonize,* according to the Laws of Language, with other Algebraical expressions?''

Hamilton himself, however, has no doubts: the study of algebra must be guided by the theoretical view, and its goal must be "to improve the *Science,* not the Art nor the Language."

These thoughts are expressed with even greater clarity in 1853 in the preface to his *Lectures on Quaternions.*[69] Hamilton again speaks of algebra as the science of pure time, asserting that it is "no mere Art, nor Language, nor *primarily* a Science of Quantity." He also takes the opportunity to recall the essay written twenty years earlier and to add this remark: "I was encouraged to entertain and publish this view, by remembering some pages in Kant's *Critique of Pure Reason,* which appeared to justify the expectation that it should be *possible* to construct, *à priori,* a Science of Time, as well as a Science of Space." Hamilton translates Kant as follows: "Time and Space are therefore two knowledge-sources, from which different synthetic knowledges can be *à priori* derived, as eminently in reference to the knowledge of space and its relations a brilliant example is given by the pure mathematics. For they are, both together [space and time], pure forms of all sensuous intuition, and make thereby synthetic positions *à priori* possible."

One should not think that this celebration of the theoretical approach led Hamilton to underrate in any way the value of applied mathematics to physics. On the contrary, Hamilton was untiring in his work as a mathematical physicist and often advised scientists not to mistrust highly abstract procedures for they too could be used in solving problems in the natural sciences. The mathematician from Dublin was, at the same time, fighting against models and the reduction of algebra to mere practice or to simple linguistics, explicitly attempting to create a new, abstract foundation for dynamics, and trying to apply the subtle formalisms of the theory of quaternions to the motion of the moon and to Fresnel's wave fronts.

We are now faced with the same type of problem as we met earlier when discussing the relation between Kelvin and Fourier: by what interpretative efforts is Tait able to accept Hamilton's approach?

If we are to judge from Tait's comments throughout the concluding pages of *An Elementary Treatise on Quaternions* of 1867, the greatest virtue of Hamiltonian methods is the evident simplicity of their applications to physical problems.[70] This praise of simplicity, however, does not have a banal meaning. As Tait writes in the preface,

It must always be remembered that Cartesian methods are mere particular cases of Quaternions, where most of the distinctive features have disappeared; and that when, in the treatment of any particular question, scalars have to be adopted, the Quaternion solution becomes identical with the Cartesian one. Nothing therefore is ever lost, though much is generally gained, by employing Quaternions in preference to ordinary methods. In fact, even when Quaternions degrade to scalars, they give the solution of the most general statement of the problem they are applied to, quite independent of any limitations as to choice of particular cöordinate axes.[71]

Tait has fully understood the advantages of the generalization of formalism that the Scottish master had always believed to be essential to the progress of mathematical science.[72] It is not surprising, therefore, that Tait was considered, and justly so, one of Hamilton's best pupils.

At the same time, however, the pupil insists on the fact that Hamilton's approach "resulted in simple, practical methods,"[73] thereby underlining the parallel between simplicity and practicality. And Tait is quite willing to emphasize the practical character of Hamilton's theories since he considers it a fundamental quality of the new calculus that it contains, as a particular case, the body of Cartesian methods, but appears particularly fruitful when its strength is tested against the problems of physics.

Simplicity and practicality: the preface to the *Elementary Treatise* gives reasons for this parallel but, to stress usefulness, completely disregards Hamilton's statements on the necessary distinction between the theoretical, practical, and philological approaches. We are told by Tait that, shortly before dying, Hamilton expressed the wish that the *Elementary Treatise* would be published as soon as possible. On that occasion Hamilton reiterated "more strongly perhaps than he had ever done before, his profound conviction of the importance of Quaternions to the progress of physical science."[74] The writing of the *Elementary Treatise* was delayed, however, due to the fact that its author was already engaged "along with Sir W. Thomson [Kelvin], in the laborious work of preparing a large Treatise on Natural Philosophy."[75] In the *Elementary Treatise* Tait intended to present both theoretical elements and concrete examples in order to induce scholars to read Hamilton's great works: *Lectures on Quaternions* of 1853, and *Elements of Quaternions*, published posthumously. But—and this sentence is quite revealing—Tait had to confess that "I have not yet read that tremendous volume completely [the *Elements*], since much of it bears on developments unconnected with physics."[76] According to Tait, then, the part of Hamilton's work that was pure research could be ignored and left to pure mathematicians to investigate; the applicative part, simple and practical, should instead be discussed and taught in universities in order to train good physicists.

Clearly, Tait's reading of Hamilton has a particular slant. The Hamiltonian approach, according to Tait, results in simple, practical methods and is thus radically different from the approaches of other mathematicians, which, "however ingenious, seem to lead at once to processes and results of fearful complexity." And so we come to the kind of indictment that will be heard again, twenty years later, in the criticism against Boltzmann. To show the difference between Hamilton and mathematicians like Warren and Argand, Tait writes in 1867 that "to a certain extent they [Warren and Argand] succeeded, but simplicity was not gained by their methods, as the terrible array of radicals in Warren's Treatise sufficiently proves."[77]

In 1867 the indictment speaks of a "terrible array of *radicals*" and in 1888 it will speak of a "terrific array of *symbols*." But this is not just a mere philological problem. It is the heart of the question. What is the genuine criterion for distinguishing between sound mathematics and "symbolic" terrorism, between a mathematics that is the idealization of common sense and arrays of symbols that betray knowledge?

As we shall soon see, the criterion derives from what is believed to be a fact and can be sketched in the following manner: a man shut up all alone and deprived of sensorial references, can reconstruct in his mind the whole of mathematics, but is absolutely incapable of explaining what becomes of a lump of sugar when immersed in water. The criterion, in short, postulates a natural gap between deduction and induction. The philosophical and scientific analysis of this criterion is at the heart of a book published by John Herschel in 1830—a book that Kelvin and Tait will use as an intermediate step in the elaboration of a Newtonianism that absorbs, without apparent contradictions, the teachings of Fourier, Hamilton, and Bacon along with those of Galileo and Laplace.

John Herschel: The Arguments of the Solitary Man and of the Lion

The argument of the solitary man and the argument of the lion, taken from John Herschel's *Preliminary Discourse on the Study of Natural Philosophy,*[78] have a history of their own, which is in part the history of a philosophical illusion.

The problem Herschel wishes to address is the following: What constitutes the "engine of discovery"? Since in our scientific investigation of the world we operate according to the intrinsic norms of an engine of discovery, it is essential for us to identify the various components of this engine if we wish to understand and promote the progress of the natural sciences. This philosophical task is extremely useful because it not only concerns the advancement of learning but bears on all aspects of social life and production. "Between the physical science and the arts of life," Herschel writes, "there subsists a constant mutual interchange of good offices, and no considerable progress can be made in the one without of necessity giving rise to corresponding steps in the other."[79] This necessary interaction is predicated on the fact that "the laws of nature, on the one hand, are invincible opponents, on the other, they are irresistible auxiliaries."[80]

Understanding the laws of nature for the purpose of enriching the "arts of life" and improving the human condition is a positive task, and its realization depends upon an accurate analysis of the engine of discovery. The first necessary condition, according to Herschel, is to eradicate a false and harmful idea about the process of discovery and to bring back the glory of Baconian method. The false and harmful idea goes back to antiquity and is condemned by Herschel with these words: "It was the radical error of the Greek philosophy to imagine that the same method which proved so eminently successful in mathematical, would be equally so in physical enquiries, and that, by setting out from a few simple and almost self-evident notions, or *axioms*, every thing could be reasoned out."[81] With the condemnation of this error goes the praise of factual observation and the warning that great care should be taken in the sciences to avoid having "some subjects particularly infested with a mixture of theory in the statement of observed facts."

We must first make a distinction between what exists in the external world and what belongs instead to the realm of thought processes, and then carefully differentiate the abstract sciences from the empirical, so that each shall be allotted its proper place in the engine of discovery. It will thus be established beyond any doubt that experience is the source of knowledge and that the life and death of every science is predicated on an "immediate and decisive *trial*" by empirical evidence.[82]

The abstract sciences operate on certain objects, and these sciences constitute a hierarchy. The set of objects, according to Herschel, includes both "those primary existences and relations which we cannot even conceive not to *be*, such as space, time, number, order, etc.," and "those artificial forms, or symbols, which thought has the power of creating for itself at pleasure." The hierarchy is given first by language and its conventional forms, then by arithmetic and algebra, and finally by logic. The student of empirical sciences must certainly possess "a certain moderate degree of acquaintance with abstract science," provided he is aware of the fact that whereas the objects of this science are well defined, the situation is very different with respect to the "words expressing natural objects and mixed relations."

Such words are equivocal: "The meaning of such term[s] is like a rainbow—every body sees a different one, and all maintain it to be the same." The meaning of a term we use to describe the world is subject to modifications, and in Herschel's view it is not easy to reconcile its various meanings, since a variation in meaning entails the veracity or falseness of the proposition containing that term. "What is worst of all, some, nay most, have two or three meanings; sufficiently distinct from each other to make a proposition true in one sense and false in another, or even false altogether; yet not distinct enough to keep us from confounding them in the process by which we arrive at it, or to enable us immediately to recognize the fallacy when led to it by a train of reasoning, each step of which we *think* we have examined and approved."[83]

The accurate study of the engine of discovery is particularly necessary in view of the fact that in the sciences we occasionally see "the triumph of theories." According to Herschel, a case in point is the triumph of Fresnel's theory. Fresnel's theory anticipated experience, gave us "a knowledge of facts contrary to received analogies drawn from an experience wrongly interpreted or overhastily generalized," and was eventually fully verified by experimental tests. Yet, Fresnel's assumptions were far removed from ordinary observation and in no way comparable to "*mere* good common sense."[84]

How is the good Newtonian to proceed? He understands the difference between the abstract and the empirical sciences and is aware of the problems that arise from the uncertainty of a word's meaning; he has also observed that in some cases theory seems to triumph over the senses and everyday experience.

How can he rationally arrive at the thesis that "the whole of natural philosophy consists entirely of a series of inductive generalizations"?[85]

According to Herschel, the first step is to examine in depth the extent of the difference between mathematics and physics, so as to understand the meaning of the tenet that experience is the only source of knowledge. Such examination must take into account the fact that the truths discussed in the abstract sciences are necessary truths, while such is not the case in the natural sciences.[86]

"A clever man, shut up alone and allowed unlimited time," Herschel writes, "might reason out for himself all the truths of mathematics, by proceeding from those simple notions of space and number of which he cannot divest himself without ceasing to think. But he could never tell, by any effort of reasoning, what would become of a lump of sugar if immersed in water or what impression would be produced on his eye by mixing the colours yellow and blue."[87]

Here is the fundamental difference between abstract reasoning by deductive steps and empirical research. The facts and laws of nature are discovered only through experience, "by which we mean, not the experience of one man only, or of one generation, but the accumulated experience of all mankind in all ages, registered in books or recorded by tradition."

The science of things is an experimental science—just like mechanics, Herschel says—and it is "a science in which any principle laid down can be subjected to immediate and decisive *trial*, and where experience does not require to be waited for."[88] If this is true, the normal behavior of the student of natural laws differs from that of the solitary man whose reasoning proceeds from the concepts of space and number. The natural philosopher must rid himself of any prejudice and of "any preconceived notion of what might or what ought to be the order of nature" and just observe what it is. In physics, in chemistry, or in mineralogy, we are dealing, not with the consequence of deductions that start from those symbols and correlations that the mind creates at its pleasure, but with purely factual questions.[89]

From these few remarks it is already clear that Kelvin's and Tait's reference to Herschel's thought is no mere homage but the result of a thoughtful reading. The senses do not deceive us if we entrust ourselves to them free from prejudices and from the temptation to prescribe with the mind what "might or . . . ought to be the order of nature," and if we subject the laws we have formulated to strict controls by means of immediate and decisive experimental tests.

Sensorial impressions are signals, Herschel writes, that pass from the objects to our minds. If we wish to comprehend nature without falling into the mistake of deducing this knowledge from the axioms of the abstract sciences, we must formulate laws and principles starting from factual questions. In contrast to the argument of the clever solitary man whose reasoning proceeds from abstract concepts, Herschel proposes the argument of the lion as an enlightening metaphor:

> In captain Head's amusing and vivid description of his journey across the Pampas of South America occurs an anecdote quite in point. His guide one day suddenly stopped him and, pointing high into the air, cried out, "A lion!" Surprised at such an exclamation, accompanied with such an act, he turned up his eyes, and with difficulty perceived, at an immense height, a flight of condors soaring in circles in a particular spot. Beneath that spot, far out of sight of himself or guide, lay the carcass of a horse, and over that carcass stood (as the guide well knew) the lion, whom the condors were eyeing with envy from their airy height. The signal of the birds was to him what the sight of the lion alone could have been to the traveller, a full assurance of its existence.[90]

Herschel finds this metaphor particularly instructive because it makes us understand that, provided we are free from prejudices, signals are always indications of some processes or operations "carried on among external objects." Thus, the fundamental problem of the engine of discovery is to prescribe such norms as will always enable us to establish rational correlations between nature's signals. The difficulties inherent in this problem are far from negligible since we have no means of knowing how far we can go in our understanding of the innermost, ultimate processes of nature in the production of phenomena.

An Engine of Scientific Discovery

The impossibility of attaining ultimate knowledge has particular relevance, in Herschel's opinion, when it bears on "the only case" seemingly free of doubts: the field of mechanics. Even in the queen of experimental sciences there is a "degree of obscurity,"[91] an awareness of the fact that motion itself "is the result of a certain inexplicable process which we are aware of, but can no way describe in words." This proves that it is unreasonable to search for the ultimate causes, and that we should aim instead at understanding the laws, ignoring the fact that we may have met a limit to our capabilities.

The cognitive problem would not be a problem, Herschel remarks, if we could "ascertain what *are* the ultimate phenomena in which all composite phenomena can be resolved." But since "there is no way to ascertain this *a priori,*" the essence of scientific research resides in an unprejudiced and strictly controlled analysis of the phenomena, without any illusions as to the likelihood of ever actually reaching the essence of things. But, once again, what are the rules of this procedure? In Herschel's words, "But, it will now be asked, how are we to proceed to analyze a composite phenomenon into simpler ones?," and, "Can general rules be given for this important process?"

"*We answer, none. . . . Such rules, could they be discovered, would include the whole of natural science.*"[92]

To avoid falling back into the fallacy of Greek philosophy, we must look directly at the signals relating to purely factual questions, formulate empirical laws, make them into "creatures of pure thought," and then reason upon them by deduction. Deduction is the "reasoning back from generals to particulars," a logical process that starts from known laws and leads to individual facts "of which we might have had no knowledge from immediate experience."

Here then is the first subdivision of the engine of discovery. First we work up from particulars to universal laws, or axioms, without having precise rules to guide us in the analysis of composite phenomena, and using mental conventions to construct those successive approximations that altogether constitute an inductive process. Then we descend by deduction from universal laws back to individual facts.

According to Herschel, the fact that the engine of discovery can be understood to operate in two distinct phases finds confirmation in the history of knowledge. Modern scientific research has been making continuous progress ever since its leaders have been the followers of Bacon and Galileo, that is, those who oppose dogmas and make a "direct appeal to the evidence of the senses" and to "experiments of the most convincing kind."[93] As followers of Bacon and Galileo, scientists like Boyle, Hooke, and Newton work on nature and represent the triumph of inductive philosophy. In Herschel's scheme, the Galilean theories that violate the senses and the Galilean experiments—the "sensible experiences"—are thus transformed into a homage to Baconian method.

Having divided the engine of discovery into two distinct parts, one problem remains unresolved, namely, the possibility that rival hypotheses may be proposed concerning specific composite phenomena. But this problem is only seemingly complex because, as Herschel writes, we can always appeal to "crucial instances" that will permit a decision "between rival hypotheses."[94] It is precisely in these instances that, according to Herschel, we perceive the necessary distinction between experiments and passive observation. "We make an experiment of the crucial kind when we form combinations, and put in action causes from which some particular one shall be deliberately excluded, and some other purposedly admitted; and by the agreement or disagreement of the resulting phenomena with those of the class under examination, we decide our judgment."[95]

It is at this point that the engine of discovery begins to appear somewhat more complex than it did after its first subdivision into an inductive and a deductive phase. The engine is not a simple mechanism of ascent and descent from the concrete to the abstract and from the abstract to the concrete, nor is this process repeated only once. When we begin the ascent we do not start from absolutely elementary and simple facts, nor do we arrive at once at laws of the utmost generality; and when we descend to the concrete it is true that we arrive at individual facts, but they are not, a priori, elementary facts that cannot be further analyzed. Herschel cannot reduce the engine to a machine that completes a definitive cycle. If it did, then the whole scheme he is attempting to outline would fall apart. The engine works at extending our knowledge and cannot stop after one ascent and one descent, for if it did, we would be faced with absolute knowledge given once and for all.

The fact that all phenomena can be further analyzed, on the one hand, and the need to appeal to crucial experiments, on the other, entail a new problem: the verification of the inductive process.

In the first place, according to Herschel, we should recognize that "almost all our principal inductions must be regarded as a series of ascents and descents, and of conclusions from a few cases, verified by trial on many." When we view the engine of discovery in this perspective, we find that it keeps grinding out information on factual matters and verifying laws and hypotheses by means of crucial experiments. The result of this inexhaustible process is what Herschel means when he speaks of science going beyond the surface of things, and of a rational knowledge that constantly arrives at "fresh branches of science more and more remote from common observation."[96]

According to Herschel, the verification or confirmation of vast and well-founded inductions is based on crucial experiments, and on having the results of induction "verified theoretically"[97] by means of particular deductive processes. In this manner we have a positive and successful scientific research that "demands continually the alternate use of both the *inductive* and *deductive* method."[98] To understand this enriched picture of rational inquiry we should discuss what Herschel calls "residual phenomena." They are exemplary instances in which we see at work deductive verification of induction as well as selection by means of crucial experiments. Predicting the characteristic properties of a residual phenomenon is done by mathematical means, or in a descending phase of reasoning. Such prediction is always bold and risky, but the situation it creates can be unambiguously resolved by the use of rigorous and decisive observations.

The example given by Herschel is a classic in the physics of the first half of the nineteenth century. It is worth recalling it to understand both what is meant by residual phenomena and in what sense mathematics is given a "certain" role in the process of discovery described by Herschel and adopted by Kelvin and Tait. It concerns the anomalies that emerge when experimental data on the speed of sound in gaseous media are compared with the theoretical predictions deduced from Newtonian theories. Newton's predictions give insufficient values and, as Herschel remarks, cannot in any way explain the "residual velocity," that is, the velocity obtained by subtracting the theoretically predicted value from the average measured values. To resolve this divergence between theory and observation, Laplace introduces into the theory corrective factors that, in far from simple mathematical forms, take into account the density of caloric and the ratio between the specific heats of the gas.[99] This enables Laplace to calculate the speed of sound taking into account the residual velocity, and thus to make a prediction confirmed by facts; all of which, Herschel remarks, constitutes a complete explanation of the residual phenomenon.

In this instance, from Herschel's point of view, we would have a deductive verification of inductively discovered principles: the verification satisfies a residual phenomenon of an anomalous nature, finds full confirmation in well-executed and decisive experiments, and constitutes an advance in physical knowledge. We extend our knowledge of the world by successive steps of ascent and descent, that is, by alternating discovery by induction with theoretical control by deduction, on the unshakable and certain basis of empirical evidence. The resulting view of the development of the physical sciences is somewhat peculiar. Laplace's intervention in the Newtonian theory of sound propagation would not appear to entail any radical change in the theory. In Herschel's view, that intervention is just a topical correction that verifies and confirms certain general assumptions.

In the more sophisticated version of the engine of discovery according to which one moves forward by small steps, each step decomposable into a succession of inductions and deductions, the ascent of thought is connected to the senses and, through signals, to the objective world; the descent, on the other hand, is regulated by abstract norms. We may then assume that in working up gradually to universal statements the scientist is not constrained by rigid rules, since nature, which is his guide, is inexhaustible and without absolute and definitive elements. In the descent, however, he has certain constraints and his activity is necessarily limited. The limitations he faces are the limitations mathematics imposes on itself in an effort to be rigorous and to operate in accordance with well-defined canons.[100]

There are no rules for the analysis of the phenomena and for starting a genuine inductive process. But there are rules, in the sense of constraints and limitations, that regulate at the deductive level the immediate verification of the gradual steps of inductive learning.

On the basis of this model of the engine of discovery, science and its history are dominated by a rigid criterion of demarcation: on the one hand, there is science, governed by Baconian rules, and, on the other, there is the confusion of presumed abstract knowledge, governed by the dogmas of deductive doctrine. Exemplary cases are for Herschel "the great laws of the planetary motions deduced by Kepler, entirely from a comparison of observations with each other, with no assistance from theory."[101]

Consequently, we should not be deceived by the rigor of theory: "Now, nothing is more common in physics than to find two, or even many, *theories* maintained as to the origin of a natural phenomenon."[102] One need only think of the theories of light and heat, Herschel remarks. On the other hand, we should not reject the whole of the theoretical approach but assess its value by the degree of its usefulness. In the first place, it is not true that there is complete freedom in *framing a theory* since the basic factors discussed in a theory or hypothesis are not chosen arbitrarily, but in such a way as to provide them with "good inductive grounds."[103] Consequently, in assessing "the value of a theory" (or hypothesis), however complex or artificial it may appear, we should not only consider its ability to describe a natural phenomenon in a satisfactory manner but determine "whether our theory truly represents *all* the facts, and includes *all* the laws, to which observation and induction lead." Even Ampère's hypothesis may be judged useful if it is able "*to predict facts before trial.*"[104]

After all, Herschel remarks, there is no danger of serious ambiguities in attempting to liberalize the empiricists' sectarian variant of New-tonianism, which forbids all hypotheses, provided, of course, we pursue the course that has been opened up by Bacon's reform. "When two theories run parallel to each other, and each explains a great many facts in common with the other, any experiment which affords a crucial instance to decide between them, or by which one or the other must fall, is of great importance."[105]

Concluding his analysis of the structure of the engine of discovery, Herschel proposes a unified view of an ever growing knowledge: "Natural philosophy is essentially united in all its departments, through all which one spirit reigns and one method of enquiry applies."[106] It is the triumph of inductive method: "Order and connexion can be traced," and knowl-edge, empirically certain, grows on "the simplicity of nature as it emerges slowly from an entangled mass of particulars."[107] And it is a knowledge that advances by a "steady, unintermitted progress."[108]

The Illusions of Empiricism

To forge a strong bond between a science that advances in a "steady and unintermitted" fashion and an objective world that is cognitively inex-haustible, Herschel establishes knowledge on forms of empirical evidence, which are always tied to induction and common sense. The resulting split between induction and deduction, and the consequent demotion of mathematics to a tool for verification, are factors that do not appear disturbing in this unified vision of the natural sciences. The price paid by the *Preliminary Discourse* is the price traditionally paid by any approach that bases rational thinking on sensorial data: the act of looking directly at nature must occur without the benefit of a pre-existing theoretical structure. The philosophical contraposition of the argument of the solitary man to the argument of the lion is a clear indication of this loss, and unequivocally declares that mathematics comes after discovery.

A similar philosophical contraposition guides Kelvin and Tait in their rereadings of scientific texts, as shown by the fact that they interpret the mathematics of W. R. Hamilton and Fourier in such a way as to reduce it to a tool of the good Newtonian. It also prompts them to rescue a portion of mathematics from the total condemnation advocated by logicians like W. Hamilton. On Kelvin's part, this rescue implies an acceptance of mathematics as the idealization of common sense; but at the same time, for both Kelvin and Tait, it implies a repudiation of mathematics as a symbolic game, as a body of naked theorems devoid of any connection with the empirical world. This philosophical contraposition is destined to have a powerful influence on large sectors of nineteenth-century culture, uniting, on the ground of the certainty of common sense, various viewpoints that are otherwise briskly opposed to one another. Here we need only mention Pierre Duhem. An implacable adversary of that English physics that, in his opinion, betrays reason in favor of fantasy, Duhem nonetheless argues that "if one doubts the certainty of common sense, the entire edifice of scientific truths will be shaken to its foundations and collapse."[109]

The historical analysis of this contraposition and of the role it played in the debates over science in the early 1800s enables us to explain certain ties between Herschel and Kelvin and better understand Tait's accusations against Boltzmann. It is significant to note that this contraposition is only apparently resolved in the field of philosophy, since the root of the problem—the relationship between mathematics and experimentation—is deeply embedded in the field of science.

From the reflections of Herschel, Kelvin, and Tait we have learned something about science and a scientist's motives, namely, that the shifting of notions and categories of objective knowledge continually undermines all interpretations and removes the absolute value of the laws, and that there is therefore a strong temptation to find in philosophy an unalterable scheme that can be adopted as an absolute system of reference. If the very idea of scientific explanation is subject to modification in the sciences, it is hard to resist the attraction of a methodology that appears to give assurance that, despite the instability and relativity of laws and explanations, there exists somewhere a sure and genuine basis for all knowledge. How, then, can one help taking a stand against those scientists, such as Boltzmann for instance, who appear to violate the very precepts that guarantee the sure and genuine basis of knowledge?

In the first half of the nineteenth century large sectors of English scientific culture are dominated by the glorification of the inductive method, of commonsense philosophy and a naive realism. However, as shown by Herschel's writings and by Hamilton's strictures against mathematicians, this glorification lends itself to considerable variations and interpretations when confronted with the fundamental problem of the existence of sciences that may be defined as abstract.

These variations are born of conflicting dictionaries in which objective knowledge and methodological metaphors meet and clash in a unique historical process of incredible richness. While the individual laws of mathematical physics appeal to relatively stable areas of the dictionary, the choice of these areas, that is, the choice of specific sectors of mathematics, is influenced in each instance by reflections arising from other areas or levels of the dictionary itself. The choice of a particular sector of mathematics is a philosophical commitment. In the case of Kelvin and Tait, we have dictionaries that reach back, through Herschel's work, to Thomas Reid's and James Beattie's analyses of the relationships between science, materialism, and skepticism in the second half of the eighteenth century.

In the eyes of the philosophers of the "Wise Club," the skepticism of D. Hume and the materialism of a Hartley or a Priestley converge, in their common recourse to real science and its problems, toward a dangerous point: the criticism of religion. Consequently, these philosophers feel they must oppose such doctrines to save both science and religion. To do this, it is necessary to set down the indubitably certain principles that are constantly being violated by skeptics and materialists. From this program is born that commonsense philosophy that, in the words of L. Geymonat, appears as a "barbarization of English empiricism."[110] Following Reid, the new Scottish philosophy maintains the existence of principles "in which the constitution of our nature leads us to believe, and which we are under a necessity to take for granted in the common concerns of life."[111] These principles cannot be proven or denied: they are fundamental. In the light of this philosophy, the skeptic is a fool for doubting the existence of the external world, and so is the materialist for denying the intuition of moral freedom: they are both in violation of the principles.

The new Scottish philosophy develops into a program comprising two different components. According to Olson's study, both a rational interest and an antirational drive can be strongly felt in commonsense philosophy. The former seeks to establish the basis of an "uncontaminated" knowledge, while the latter denies that "reason alone could provide the basis for a significant knowledge either of the material or of the spiritual world."[112] One must heed Bacon's teachings on the inductive method and Newton's warning never to proceed by hypothetical reasoning. It will thus be clear that empirical truth is found only by those who conform to the norms of common sense.

Although the difference between mathematical and physical knowledge is never to be forgotten, Bacon's doctrine must be sufficiently liberalized to incorporate mathematics' splendid edifice. However, geometrical objects and mathematical concepts appear to Reid as disquieting elements. Geometrical objects, as Reid writes in a letter to James Gregory,[113] may be "possible modifications of things which we daily perceive by our senses." And by a careful analysis of the sense of touch a correlation might perhaps be found between mathematical concepts and the properties of bodies. Everything must be ultimately brought back to the senses and to the certainty of common sense. A century later, Kelvin still remembers that Reid has shown the way to a knowledge based on what our senses teach us. [114]

Reid's program is inconclusive and generates variants that, although at times antithetical, leave untouched both the primacy of the senses and a superficial view of the distinction between induction and deduction. Since god has created the world according to laws and has endowed man with an inherent drive to discover them by looking at facts and gathering them in ever more general forms, and since the principles of common sense are inviolable, induction and common sense coexist harmoniously in the good Newtonian, who has no reason to doubt either the world or religion. However, these beliefs cannot uniquely define the function of mathematics and geometry. William Hamilton proclaims that "the mathematician knows nothing"; John Leslie considers mathematics "an artificial mode of procedure"; Dugald Stewart makes a distinction between geometry—a science closely tied to things—and analysis, which is continually vitiated by paradoxical and absurd conclusions;[115] Herschel endeavors to save deduction by reducing it to an act of immediate verification of the inductive process; Kelvin states that mathematics is splendid when it represents the idealization of common sense; and Tait, on his part, erects a barrier between sound deduction and mathematical terrorism.

Under the dark shadow cast on English physics for over a century by this philosophical metaphor, a war is brewing—the war that Tait will wage against Boltzmann in the name of the realism that should characterize the good Newtonians. In vain will Boltzmann remind his enemies that theory conquers the world and that we cannot express a single sentence that translates into reality a pure fact of experience.[116] Even though at the end of the century Kelvin's program ends in failure, the illusions of eclectic empiricism remain untouched and the metaphor of Herschel's lion survives the triumphs of theory.

Boltzmann's Philosophical Battle

From 1866 to 1884, Boltzmann's mathematical theories of molecular phenomena open up unsuspected views in nineteenth-century theoretical physics. The power of algorithms takes on disturbing connotations and in turn casts strange reflections on the as yet unexplored network of interactions that appears to connect mechanics, the theory of gases and of radiation, electromagnetism, and the calculus of probability. Most empirical data seem to rebel against Boltzmann's results and large sectors of the philosophy of science ridicule the Viennese physicist's attempts to analyze the inner properties of the microuniverse by means of reason and symbols.

As for Boltzmann, isolated in a paper world populated by irreverent conjectures and by arrays of algebraic symbols, he strives for decades to defend the virtues of theoretical work against the supremacy of the senses. The problems of physics, he says in 1904—two years before committing suicide—cannot be resolved by the philosophical act of "summoning [empirical] data to the judgment throne of our laws of thought"; what is rational and necessary, instead, is "to adapt our thoughts, ideas and concepts to what is given."

In support of theory, Boltzmann defends himself from Herschel's lion by attacking Ostwald's superficial philosophy, rereading Mach, mocking the dogmas of certain schools of atomism, and conceiving of the progress of physical science as a nonlinear development of rationality.

Even in this partial reconstruction of some weak areas of Boltzmann's dictionary we glimpse a picture that had already started to take shape in the preceding analysis. As the network of connections becomes increasingly clearer, we find confirmation of the fact that the autonomy of objective knowledge is a virtue that cannot be attained without a hard struggle. Scientific progress cannot be measured by piling law upon law, but by assessing the evolution of knowledge and analyzing the logic that governs each step of scientific research. And this evolution drags along philosophy as well.

4   *In Praise of Theory*

Theory and Technology

"The more abstract the theoretical investigation, the more powerful it
becomes," Boltzmann says in 1890.[1] Against the followers of those
philosophies that presume to evaluate the sciences purely from a practical
standpoint and to reduce them to techniques devoid of any knowledge,
Boltzmann argues that the Eiffel Tower or the Brooklyn Bridge could
hardly have been built without a mathematical theory of elasticity. That
tower and that bridge rest not only on solid iron structures, but, most
important, on an even more solid mathematical theory. And this theory
is not only useful for building sound structures, but is part of that body of
notions, principles, and laws that the power of abstraction organizes into
an ever more rational and coherent view of the external world: theory,
Boltzmann writes, conquers the world.

The progress of knowledge is such, Boltzmann remarks with heavy irony,
that "even those who value theory only as a milk cow, can no longer doubt
its power."

There is certainly a widespread attitude of contempt toward abstraction.
"A friend of mine has defined the practical man as one who understands
nothing of theory and the theorctician as an enthusiast who understands
nothing at all." However, Boltzmann remarks, if it is true that we must
take care not to mistake mathematical correlations for what actually
exists, it is equally true that we must steadfastly refute the poet's saying
that theories are much too drab in comparison with the richness of life.
And if it is true that this poetic lament comes from Goethe, it is also true
that Goethe himself has the devil say, "Do but spurn reason and science
. . . you will be unconditionally mine!" Boltzmann, in sum, takes inspira-
tion from many sources when he asks the question, "What is theory?,"
and finds complex answers for it. In the first place, theory's task is to
create an image of the external world; in the second, theory must be aware
of the fact that the image is distinct from the external world; finally,
theory must be "our guiding star in all thought and experiment."

The defense of theory against the temptations of the irrational and the brutal transformation of science into technology is indeed the dominant theme in Boltzmann's thought. When he adopts Riemann's arguments that objective knowledge becomes the more rigorous and deep the further removed it is from the surface of things, or, in his own words, that the more abstract the investigation the more powerful it becomes, Boltzmann is in fact restating in a modern key Galileo's axiom that theories violate the senses. This re-elaboration, solidly based on the principles of mathematical physics, clashes with the philosophies of Boltzmann's many critics. At the end of the nineteenth century there are powerful stimuli and incentives for philosophy to declare a crisis in scientific thought, and the declaration of such a crisis demands the defeat of all those who believe themselves to be the witnesses of a scientific revolution instead.

Attacked by the English physics of Kelvin and Tait in the field of applied mathematics and isolated among the atomists who see in his theories inconvenient excesses of generalization and abstraction, Boltzmann will in fact be defeated on philosophical grounds. And his defeat is decreed by majority rule. At the 1895 Congress of Lübeck the view of the world and of scientific theory upheld by Boltzmann is rejected by a block of scientist-philosophers led by Mach and Ostwald. It would be a mistake to consider that meeting as an isolated event of little importance. That meeting and that majority vote reflect a vast and complex historical process in which the philosophical controversy is dominated by intellectual persuasions determined to declare the bankruptcy of objective knowledge and to impose the supremacy of different forms of knowledge upon the sciences.

After considering this particular aspect of Boltzmann's defeat, it is undeniable that some parts of Boltzmann's thought are of considerable interest even today, when we see the hostility to reason re-emerging in seemingly new forms.

Mechanical Explanation and Objective Problems

The same year that Tait launches his attack against the physicists who replace thought with algebraic terrorism, Boltzmann presents at the Imperial Academy of Sciences a particularly unpopular version of the right approach to scientific investigation.[2] It is not true, the Austrian physicist argues, that in the study of nature the most direct path is also the best. "The most direct path would be to start from our immediate observations and to show how by means of them we attained knowledge of the universe." But in the natural sciences we must pursue instead the opposite course. Sensations should not have any bearing on scientific matters, as we have learned from the fact that Copernicus' contemporaries "felt" that the Earth did not rotate.

Instead of starting from sensations, science should silence them. "I have silenced feeling: if the hypothesis explains all the phenomena concerned, feeling will have to give way as in the question of the earth's rotation."

This is a classic Galilean argument. But how can it be used when, in the wake of Herschel, Galileo has been enlisted in the army of the good Baconians and experience is upheld as the only source of knowledge? The problem, then, is to make a distinction between experience and sensations. In experimenting, the scientist is guided by theory, which furnishes rules and indications for the definition of a problem. Sensations, instead, are something that concerns the nervous system.

A scientific experiment is a project that narrows down the field of research in order to investigate it rigorously, and in this sense the most direct path is by no means the best for the natural sciences to follow in their attack on the world. To clarify his idea, Boltzmann uses the following metaphor:

If a general intends to conquer a hostile city, he will not consult his map for the shortest road leading there; rather he will be found to make the most various detours, and every hamlet, even if quite off the path, will become a valuable point of leverage for him, if only he can take it; impregnable places will be isolated. Likewise, the scientist asks not what are the currently most important questions, but, "Which are at present solvable?", or sometimes simply, "In which can we make some small but genuine advance?" As long as the alchemists merely sought the philosopher's stone and aimed at finding the art of making gold, all their endeavors were fruitless; it was only when people restricted themselves to seemingly less valuable questions that they created chemistry. Thus natural science appears completely to lose from sight the large and general questions; but all the more splendid is the success when, groping in the thicket of special questions, we suddenly find a small opening that allows a hitherto undreamt of outlook on the whole.

The road to knowledge, in sum, does not start from sensations and is not a straight line but a roundabout course through topical problems in which each move is guided by sets of rules and hypotheses. The general laws are the ultimate goal of this tortuous journey through topical problems amenable to solution. In short, the discovery of the laws is at the end of the journey, not at the beginning.

This is the second dominant theme in Boltzmann's thought. The praise of theory is always accompanied by the awareness that the approach to research is not linear, the splendor of reason being heightened rather than dimmed by immediately verifiable procedures whose historical development defies continuity. Though pursuing a roundabout course, objective knowledge will never lose its way in the thicket of solvable problems. On the contrary, it is only by pursuing a roundabout course that the general problems can be surrounded, then circumscribed within a previously explored and defined context, and finally conquered.

The strategy of attacking specific problems is of course facilitated by the specialization of scientific work. But at the same time, "the various disciplines are so increasingly tied to one another that, despite the extreme division of work, nobody should ever lose sight of the other fields," although it may be impossible to enter into the details of those branches of scientific inquiry in which one is not a specialist. The overall unity of the enterprise is not a philosophical myth. Thus, if he wishes to be a true specialist, the individual scientist must have some knowledge of what other scientists are doing in different disciplines.

Boltzmann's view of science can now be summarized as follows. Knowledge is attained through ever more abstract theories. These theories develop by violating the senses and by proceeding on a roundabout course through topical solvable problems. Theories guide experience and are all part of the same rational enterprise. The construction of the physical world—intended as a theoretical image of the external world—not only obeys requirements of a practical nature, but, as a strategy aimed at attaining knowledge of nature, also proceeds within human history.

It may now be asked whether and to what extent this is a mechanistic view. At the 1886 meeting, Boltzmann states that the nineteenth century will be remembered not only as the century of iron, steam, and electricity, but most of all as the century of the "mechanical view of nature" and of Darwin's evolutionary theory. How are we to understand the "mechanical view of nature"?

Two seemingly distinct answers may be given to this question. The first answer is that we must define precisely what Boltzmann means by mechanical explanation. In 1897, in a paper devoted to the defense of atomism, Boltzmann expresses his support for the epistemological battle against "those framers of hypotheses who hope to find without effort a hypothesis that would explain the whole of nature," which is a legitimate battle against "the metaphysical and dogmatic foundations of atomism "[3] Given this premise, it is entirely proper to discuss the question of mechanical explanation in a nondogmatic manner. "If by a mechanical explanation of nature we understand one that rests on the basis of current mechanics, we must declare it as quite uncertain whether the atomism of the future will be a mechanical explanation of nature." Boltzmann is quite clear on this point: if atomistic physics is to establish in the simplest possible way the laws of "temporal change of many individual objects in a manifold of probably three dimensions," then it may be said that it is "a mechanical theory, at least in a metaphorical sense." Hence designating a law of temporal change as mechanical or nonmechanical "will be entirely a matter of taste."[4]

The second answer is based specifically on Boltzmann's mathematical physics. In his early writings on the kinetic theory of gases, Boltzmann's approach was clearly reductionist, in the sense that he planned the mathematization of the theory with the explicit intent of reducing it to a chapter of rational mechanics. But in the papers of 1871 we already see a significant change in his approach. At the start, Boltzmann's intent was to elaborate a "purely analytical proof" of the second law of thermodynamics, and in 1866 he claimed he had transformed that law into an "axiom of pure mechanics," since, in his opinion, it could be legitimately maintained that the law in question corresponded to the principle of least action.[5] In 1871, however, Boltzmann revised his work by introducing into the theory, in increasingly more radical forms, notions and inferences derived from the calculus of probability.[6] This revision resulted in 1872 in a monumental monograph that went far beyond his own previous research and the studies carried out in the same years by Maxwell, Clausius, and Szily.[7] The problem was no longer to find deductive chains that could eliminate the exceptional position of the second law by bringing it back to mechanics, but to explain the second law through an analysis of its relationship to the calculus of probability. While enunciating a fundamental theorem on the subject (the E theorem, or first statement of the H theorem), Boltzmann correctly remarked, "We are on the way to an analytical proof of the second law following an approach entirely different from any hitherto attempted."[8] A few years later, in a second monograph on the subject, he stated even more explicitly that the problem consisted in "finding the relationship between probabilistic propositions and the second theorem of the mechanical theory of heat."[9]

This internal shift in his program of theoretical research was thus tied to the new correlations that emerged in the construction of the theory, in the sense that the latter was undergoing gradual modifications contingent upon a specific mathematics. The probabilistic propositions could in no way be reduced to a translation into new symbols of deductive chains previously established; in effect, they were bringing forth new concepts that had to be further analyzed and that, by their very nature, stimulated further developments and changes in the original theoretical apparatus.

Was it still a mechanical explanation? The question was inevitable, since doubts were being raised in many quarters about the cognitive power of the calculus of probability. Maxwell, for instance, although himself using probabilistic notions in his research on the theory of gases, was convinced that the application of such notions meant giving up the exact knowledge that only mechanics could afford.[10] Boltzmann, on the other hand, saw no such loss. "One must not confuse a proposition not fully demonstrated, and whose precision is consequently problematic, with a proposition of the calculus of probability, which is rigorously proven: the latter represents a necessary consequence of specific premises."[11] The probabilistic computation of molecular states and the mathematical consequences derivable from it represented for Boltzmann the only possible way to formulate an exact theory. The question of "mechanical explanation," in sum, had become the open question of the relationship between probability and the second law and could no longer be considered in terms of a mere reduction of the second law to mechanics. In a general way, therefore, it is quite understandable that years later the designation of an atomistic or probabilistic theory as mechanical or nonmechanical would appear to Boltzmann simply as a matter of personal taste or of linguistic metaphors.

It should now be clear why the two answers to the initial question—What is meant by a mechanical view of nature?—are only apparently different. In Boltzmann's thought, the definitions of "mechanical explanation" and of "mechanical view of nature" are drawn from the whole of his dictionary, although from time to time, in enunciating those definitions, general considerations prevail, as well as opinions closely tied to the particular areas of rules that Boltzmann himself is emphasizing in his dictionary. On the other hand, the dictionary is historically determined and changes with time. Thus Boltzmann's answers vary depending on the shifts occurring in his dictionary, which reflect both the internal modifications of the kinetic theory and the more specifically epistemological changes caused by the cultural battle Boltzmann is waging. Consequently, the problem is not to draw, once and for all, a logical demarcation line within Boltzmann's dictionary between his philosophy and his mathematical physics. Rather, the problem is to examine the specific relationships that are historically established between different areas of his dictionary, taking into account the fact that the boundary lines between such areas are never rigidly drawn.

On What Exists

In his praise of theory, a point that Boltzmann always emphasizes is theory's power to conquer the "external world." Such a statement implies a belief in the objective existence of the processes of inanimate matter and therefore in preconditions for knowledge. These issues are not ignored by the Viennese physicist, who confronts them head on in a paper of 1897[12] —the same year in which he also defines the meaning of the concept of mechanical explanation.

The controversy with Ostwald and with the advocates of phenomenalistic physics, a constant reference point in Boltzmann's analysis, makes it even more necessary to clarify the motives that may influence a physicist's choice between realism and idealism at a time when many proclaim the disappearance of matter and view theories as purely utilitarian correlations devoid of any objective knowledge.

We should hasten to add that Boltzmann does not include a demonstration of the existence of the world among the factors that may influence the choice between realism and idealism. On the contrary, he states that *"it would of course be absurd to prove or disprove the objective existence of matter."* According to Boltzmann, the real problem is to define what is meant by basis or precondition for scientific knowledge. Since this question already entails serious difficulties, it will be opportune to begin a consideration of preconditions by recalling the basic framework of Euclid's geometry: "Just as in geometry Euclid begins with unprovable axioms, we shall begin by examining what facts constitute the basis and precondition for knowledge."

An essential precondition consists of the regularities found in sense perceptions and in the impulses of the will. These regularities can be enunciated. A case in point is the law of causality, which, as Boltzmann remarks, "we are thus free to denote as the precondition for all experience or as itself an experience we have in conjunction with every other." At this level we form impressions, memories, and images of the world as well as expectations. The surprise produced by unexpected events is later overcome by means of further corrections and readjustments of the image, in a process that may be rendered ever more complex in an infinite number of ways and that leads to the formulation of opinions.

We have not come to theories yet. But at this level we already have organized forms of remarkable complexity, even though the experiences with which we test images and opinions are not scientific. The opinions that are consistently confirmed by experience and appear certain become part of our knowledge.

At a certain level of complexity concepts come into play. "To construct thought-pictures we constantly need designations for what is common to various groups of phenomena, thought-pictures or intellectual operations: we call them concepts."

Boltzmann is aware that this interpretation, based as it is on preconditions contingent upon sense perceptions, may open the way to solipsistic considerations. How can we be certain that the regularities perceived by one individual coincide with those perceived by others? In building thought-pictures we use signs and rules. How can we be sure that these signs and rules hold for others too?

Once again, Boltzmann takes the position that he will not ask for proofs of existence or nonexistence. "If, however, someone were to assert that only his sensations exist, whereas those of all others were merely the expression in his mind of certain equations between certain of his sensations (let us call him an idealist), we should first have to ask him what sense he gives to this, and whether he expresses that sense in the appropriate way." From this point of view, any statement of existence or nonexistence does not require a proof but implies a reference to rules. The question of the existence of a planet, or of other human beings, cannot be decided by demonstration. There has been a long debate, for instance, on the possible existence of a planet, as yet undiscovered, which might be responsible for the perturbations observed in Mercury's orbit. The reasoning that leads to conjectures on the existence of this planet, Vulcan, is the same that earlier led scientists to infer the existence of the planet Neptune from the perturbations in Uranus' orbit. Neptune's orbit was then computed and finally the new planet was actually observed. But this has not happened in the case of Vulcan. While discussing this case, Boltzmann notes that any question about the existence of the unicorn or of the planet Vulcan has a precise meaning only if we understand the term "to exist" in the same sense as we understand that the deer or the planet Mars exist: this sense, according to Boltzmann, is clarified by the "empirically known relation to the second two items."

Having concluded that ontological problems cannot be solved by demonstration and that the appropriate way to discuss them is in accordance to rules, Boltzmann feels justified in strongly criticizing the position of those who proclaim the disappearance of matter. He remarks, "Since we have reserved the term 'not to exist' for the satellite of Venus, the philosopher's stone and so on, it would evidently be inappropriate to say that matter 'does not exist.' "

Boltzmann, who is far from being a vulgar materialist, notes at this point that the foregoing discussion has simply defined "the concept of subjective existence or nonexistence." Now, is it possible to define what we really mean when we say that "we must adopt the objective point of view"?

"If therefore I am to make myself understood," Boltzmann writes, "I must adopt a language in which all things exist on the same footing ('objectively'). This adherence to the language of others, which is given to me in experience (because learnt), I call the objective point of view, in contrast to the subjective one so far described." Thus, "If somebody regards it as obvious *a priori* that matter does or does not exist, this, in the absence of some prejudice, can be considered only as expressing the subjective conviction that either one or the other designation would lead to quite ludicrous complications.".

The starting point of Boltzmann's argument is a desire to elucidate the meaning of the thesis that the sensations of other people "do not exist." If the attribution of "existence" and "nonexistence" is determined by the same rules that apply to the existence of the deer or the planet Mars, then that thesis becomes untenable, and at least two significant consequences ensue. In the first place, nothing dramatic happens to matter when we decide, on the basis of rules, that the unicorn or the planet Vulcan do not exist. In the second place, we are entitled to state that no line can be drawn separating the processes of inanimate matter from psychological processes. According to Boltzmann, this in turn implies that it cannot be rationally maintained that the former "do not exist" and only the latter "exist."

In effect, he remarks, it would not be appropriate to deny the existence of sensations in animals that we classify below a certain level of organization. Hence, proceeding by successive steps, it would become untenable to deny the existence of unorganized, inanimate matter.

Since we have now established some concepts, "the denotation must always be so chosen that we can operate with the same concepts in the same way under all circumstances, just as the mathematician defines negative or fractional exponents in such a way that he can operate with them as with integral ones."

Those who follow different approaches are bound to meet with unnecessary difficulties. Boltzmann tells how somebody once attempted to demonstrate that, since a high school teacher is really a professor, therefore our law is the only right one because it gives the teacher the title of professor. "I have the same feeling," Boltzmann writes, "when a word like 'exist' is singled out from the language and, without fixing its sense, people start racking their brains as to what exists and what does not."

Boltzmann goes on to remark that progress in thinking will be furthered by the elimination of all those modes of reasoning that embroil rationality in contradictions. We must eliminate those "forms of inference and concepts [that] always arise when originally appropriate modes of thought are transferred to cases where they do not fit." We must progressively readjust our thoughts and define the meaning of words instead of using them uncritically in ambiguous situations.

Let us go a step further. Instead of separating with an insuperable barrier the processes that occur in inanimate matter from thought processes—as the proponents of phenomenalistic physics wish to do—let us eliminate this distinction on the ground that thought processes are a particular case of natural processes in general; this will permit the construction of "a simpler (objective) world picture." This additional step, suggested by Boltzmann in a context in which the brain is regarded as an apparatus, or an organ, that produces word images, is particularly important for defining better the materialistic assumptions of Boltzmann's realism as well as the meaning of mechanical explanation within this realism. Boltzmann speaks of a mechanism that develops in the mind of man and then hastens to note, "The term mechanism is of course not meant to prejudge whether the laws of current mechanics must suffice to represent it." This point is further clarified in a subsequent note in which Boltzmann discusses the laws of analytical mechanics and states that "we are by no means sure that the whole of inanimate nature can be represented by these latter."[13]

For Boltzmann, then, thought processes are specific material processes occurring in the brain. And this, in his opinion, constitutes the realistic point of view. There is a different point of view, however, whose supporters place thought processes above those that occur in inanimate nature. But, once again, Boltzmann's position is that there is no problem whose solution will prove either view true or false. The only question is to determine the appropriateness of each view to a particular objective. As far as progress in thinking and the construction of images of the external world are concerned, we have seen that Boltzmann emphasizes the need to eliminate the contradictions that arise when certain inferences and concepts, suited to the specific conditions that produced them, are transferred to other fields of inquiry. In these cases, which are peculiar to scientific research, the realistic approach appears to be far preferable to the idealistic one, although there is no way to demonstrate it.

The elimination of contradictions is necessary, then, but contradictions arise only in the context of inferences and concepts. As Boltzmann remarks, "Contradictions . . . can lie only in ways of denoting and are thus a sign that these have been inappropriately chosen." Contradictions do not concern experience: "Experience cannot contradict itself, for even if its laws were to change completely, ways of denoting would have to adapt to the new laws."[14] With these considerations Boltzmann is far from proposing anew a rift between the realm of theory and that of experience. In the first place, he always insists on the fact that experiments are projects guided by theory and that, to arrive at knowledge, the path of research should not start from sensorial data. In the second place, he opposes just as strongly the idea that "sensations are simple, or qualitatively different from processes in inanimate nature." Who would know how to draw a line between the two realms? The realistic point of view, which has accepted the impossibility of separating the psychic from the inanimate, is therefore much more appropriate than the idealistic view.

"If one adheres to this our view," Boltzmann writes, "processes in inanimate nature differ so little in quality from animate ones, that it is impossible to draw any boundary, and it would be impracticable to ascribe objective existence only to sensations but not to processes in inanimate nature."

More important, by adhering to this viewpoint we obtain a general result of the utmost significance. There are questions that have long been waiting for answers and have to do with matter and spirit, existence and nonexistence, and whether atoms are endowed with sensations. The lack of answers is due to the fact that whenever one attempts an answer one is faced with contradictions. But the contradictions disappear, Boltzmann concludes, as soon as "one realizes that one did not know what one was really asking." One realizes, in short, that the dispute was about problems that cannot be solved because they are badly stated.

On the Method of Theoretical Physics

Let us return to the methods of theoretical physics and to the question of the division of scientific work. Toward the end of the century Boltzmann reopens the discussion on the two aspects of the division of scientific work: the positive aspect, due to specialization in research, which furthers the growth of physical knowledge; and the negative aspect, represented by the danger of losing sight of the all-inclusive character of knowledge itself.[15]

The danger is not only that science will be split into separate disciplines. In Boltzmann's view, it is a danger that affects research directly, by obstructing it from within in both the discovery of the new and the restructuring of the old.

To overcome this danger, according to Boltzmann, we should first of all become aware of a particular characteristic of modern theoretical science, namely, the presence of a strong tie between the method of theoretical physics and the theory of knowledge.

Modern theoretical physics has for Boltzmann the following virtue: while reflecting on things, we also reflect on the method by which we think. Hence the importance of the theory of knowledge to science. Its importance should not be overshadowed by the foibles of old metaphysics, but rather emphasized by historical reconstruction and by the elimination of philosophical pseudoproblems.

If we regard the development of theory from a historical perspective we soon realize that "it is by no means as continuous as one might expect, but full of breaks and at least apparently not along the shortest logical path."

Even a brief glance at the development of modern theoretical physics cannot fail to reveal the influence that Galileo's and Newton's theories have had on it. For many decades the theoretical precepts of mechanism appeared as the sure bases for all knowledge and along with atomistic theories guided research toward a specific explicative goal. "The task of physics," Boltzmann writes, "seemed confined forever to ascertaining the law of action of the force acting at a distance between two atoms and then to integrating the equations that followed from all their interactions under appropriate initial conditions."

However, these developments belong to the past. As Boltzmann remarks, they now appear as monuments of ancient scientific memories. Nevertheless, it is extremely important to defend the old doctrines and to elaborate the results of "classical theory" in a clear and logically ordered manner. In taking its defense upon himself alone, Boltzmann remarks, "I therefore present myself to you as a reactionary."

One may wonder whether this ostensibly reactionary posture means that Boltzmann, after considering the problems of the theory of knowledge and of a history of physics that does not follow the shortest logical path, is simply falling back into the fold of mechanistic philosophy. Before we answer this question, let us look at the arguments advanced by Boltzmann himself.

In proclaiming himself a reactionary, Boltzmann by no means proposes to engage in an all-out defense of the mechanistic dream. Rather, he intends to defend two general theses concerning rationality. In the first place, according to Boltzmann, we must resist all methodological temptations to denote certain ways of doing theoretical physics as the sure foundations of absolute knowledge, whether these ways claim to be founded on mechanical explanation, or whether they seek to base knowledge securely in energetics or in phenomenalistic physics. In the second place, Boltzmann reminds physicists and philosophers that only half of our experience is ever experience: experience is always intermixed with theory. By boldly overcoming mere empirical evidence we discover ever more surprising facts, even though in the attempt we may run the risk of making mistakes. But these characteristics are a virtue of scientific inquiry, not the source of its shortcomings.

Expanding on these two general theses, Boltzmann does not confine his reflections to a dutiful defense of old theories: on the contrary, he develops a stringent critique of phenomenalism and energetics and at the same time advocates the need for a thorough revision of the molecular approach as a rational basis for theoretical research.

In his schematic reconstruction of the nineteenth-century scientific revolution, Boltzmann notes that the first serious attack on the old theories was directed at Weber's electrodynamics, and that this attack developed simultaneously along two fronts, one physical and one epistemological.

On the epistemological front, Maxwell demonstrated the fallacy of all theories claiming to offer a definitive interpretation of electromagnetic phenomena or an absolute explanation of the true nature of matter. On the physical front, it was Maxwell again who formulated a mathematical theory opposed to that of Weber, replacing action at a distance with action by contact. And while the old theories were not capable of predicting such phenomena as Hertz's oscillations, the new theory could do so.

On the strength of these undeniable differences, Boltzmann writes, "Extremists went so far as to brand all conceptions of classical physical theory as misguided." Yet, the enormous advances made in the field of radiation have characteristics that cannot be decided by crucial experiments. Consequently, we should exercise great caution before concluding that experimental verification of a theoretical consequence is proof of the absolute correctness of the theory. A theory can never be judged in terms of absolute truth or absolute falseness. The function of laboratory experiments is not to provide the elements for a definitive judgment but to show that the field of research is unlimited.

It is certainly true, Boltzmann remarks, that "theory . . . has been shaken from the complacency in which it thought it had already recognized everything." But it is also true that the new phenomena have not yet been inserted in a new, unified theoretical framework and that research in the field of mathematical physics is in a state of indecision and ferment.

According to the Viennese physicist, this is particularly evident when we examine the *status* of mechanics. The reform made by Kirchhoff was only a formal one. While, on the one hand, this reform succeeded in sheltering mechanics from obscure metaphysical disputes on the nature of force, on the other, since it did not introduce substantial modifications into classical mechanics, it retained a purely formal character. Hertz went a step further by analyzing the concept of mass and outlining interesting lines of research for the future. However, Boltzmann remarks, Hertz has left completely unresolved a number of perplexing questions about the notion of hidden mass. In the current state of theory, hidden mass is not just something hidden but is above all something we cannot clearly conceive.[16]

Thus, Hertz's program remains a program for the distant future. However, it has made physicists aware of the mistake that Maxwell had earlier recognized in the cognitive pretensions of Weber's approach: the old theories are in error at the epistemological level because they equate the objectivity of knowledge in physics with the absolute representation of nature. We know today, Boltzmann remarks, that it is not in our power to find an absolutely correct theory.

This new awareness has provoked amply justified reactions against those who have constantly advocated the need to build models along the lines indicated by classical theoretical physics that by means of either atoms or vortices would explain god's design once and for all. But these legitimate reactions have also brought about rifts among physicists, with the result that many have fallen back onto the old metaphysics that one thought had been completely discredited.

Boltzmann identifies two groups of dissenters: the extremists, who have promoted energy to a dogma and never tire of proclaiming the disappearance of matter; and the more moderate ones, who uphold the glories of phenomenalistic physics.

In keeping with the scathing criticism voiced by Planck a few years earlier,[17] Boltzmann writes that the energeticists embroil themselves in superficial analogies, unsubstantiated inferences, and outdated philosophical disputes.

The moderate dissenters are in turn divided into two parties: mathematical phenomenalists and general phenomenalists. The former confine mathematical theory to practice alone. In their opinion, theories consist of two parts, one certain and one modifiable, and the criteria for choosing between the two rest solely with the practical use of the equations. The latter reduce theories to mere descriptions of phenomena or reviews of their natural history. By giving up as false any form of mechanical explanation, they also give up any basis for rationality and any hope for a unified view of nature.

The two groups of moderate dissenters have one point in common: both assert that science is science insofar as it represents the phenomena without outreaching experience.

And therein, according to Boltzmann, lies their illusion, not their strength. Only half of our experience is ever and solely experience:

> Phenomenology believed that it could represent nature without in any way going beyond experience, but I think this is an illusion. No equation represents any processes with absolute accuracy, but always idealizes them, emphasizing common features and neglecting what is different and thus going beyond experience. That this is necessary if we are to have any ideas at all that allow us to predict something in the future, follows from the nature of the intellectual process itself, consisting as it does in adding something to experience and creating a mental picture that is not experience and therefore can represent many experiences.

Boltzmann reminds all dissenters that the defense of the old physical theories is concerned not so much with saving the dogma of an absolute knowledge of real things, as with supporting the rational foundation of an objective knowledge that is capable of change. The approach of molecular physics is not dogmatic, therefore, to the extent that it is aware of the illusion of the absolute without, however, falling into the philosophical extremism that denies any form of knowledge of the real world.

Turning from the old philosophy to the new, Boltzmann argues that the greatest merit of the theory that explains the relationship between entropy and the calculus of probability is not that it furnishes a definitive explanation. Its merit lies in its ability to widen the horizon of scientific inquiry and to suggest new relationships between concepts and novel experimental circumstances. Molecular physics is undoubtedly beset with difficulties. Just to mention one, there is the problem of the discrepancies between theoretical predictions and laboratory data in connection with specific heats. But these discrepancies are not tombstones to be set upon an old and outmoded theory, but rather an incentive to deepen our understanding of the structure of molecules. It is not a question of destroying a faulty theory but of understanding that that theory shows molecules to be vastly more complex than anyone could have imagined. This problem cannot be solved either by denying that molecules "exist," or by stating that only energy exists or that physics describes phenomena.

Mechanics and Mechanism in Boltzmann

In 1900 and in 1902 Boltzmann takes up again some of the themes we have just analyzed and discusses them in two inaugural lectures intended mainly for students.[18] In a brief foreword to the published text the author writes that he will perhaps cause some disappointment to the readers who expect to find in his lectures a re-elaboration of the philosophical criticism he had previously expressed. The reason for this is very simple, Boltzmann explains: philosophical criticism is necessary when the audience is made up of scholars surfeited with science and therefore quite willing to take "the digestive tablets of critical philosophy." The situation is entirely different when one lectures to young people who wish "first to absorb scientific theories rather than give them out again."

In the lecture of 1900, having set aside the digestive tablets of critical philosophy, Boltzmann concentrates on extolling the importance of analytical mechanics. Analytical mechanics is the gateway to all of theoretical physics and the concept of mechanism is the key to the understanding of every existing thing. The search for a regulatory mechanism[19] is the right way to untangle the knots present in natural phenomena, in politics, in social life, and in the realm of values.

While trying to communicate to the students his enthusiasm for analytical mechanics, Boltzmann also invites them to reflect on the latest developments in theoretical physics and to ask questions such as the following: Is mechanical theory too "mechanical"? If mechanics means obedience to the rules, Boltzmann answers, then no theory is too mechanical. But if mechanics implies the temptation to equate theoretical concepts with things in the external world, then it must again be pointed out that there cannot be any coincidence. Although the mechanical interpretation has dominated a number of disciplines, it has been severely criticized in its main field of interest, namely, theoretical physics. Its pillars have been shaken, Boltzmann remarks, and its fundamental principles have been found to contain obscurities. At present, any attempt to demonstrate a priori that every variation should be reduced to a motion of parts is necessarily based on metaphysical arguments that are totally inadequate. Suffice it to say, Boltzmann notes, that the electromagnetic theory is being discussed as a possible substitute for the mechanical theory.

This reversal of roles between mechanics and electromagnetism is still present in the lecture of 1902. But, once again, Boltzmann's contention that analytical mechanics is the heart of theoretical physics is not sufficient reason for branding him as a mechanist. In presenting analytical mechanics as the vital center of all theoretical physics, Boltzmann by no means defends a naive, absolute vision of the external world; rather, as it should be clear by now from all we have said, he defends the rationality of a theoretical approach that describes the world objectively and consists of "intricate theorems, ultrarefined concepts, and complicated proofs." Boltzmann's point is that mechanics should not be taught as a doctrine absolutely valid according to the standards of questionable metaphysical principles. One should teach instead the Hamilton–Jacobi theory and explain that its results are valid not only because they enable us to manufacture objects, but because they allow us to build, not necessarily by the shortest logical path, an image of what we call the external world that is well reasoned and organized according to verifiable norms.

Following this train of thought, Boltzmann writes in 1903 that Mach is quite right in saying that no theory is ever completely true or completely false and right again in asserting that a theory thrives by being subjected to violent attacks.[20] However, he cannot refrain from taking Mach to task for a statement that is essentially a philosophical thesis of a general character: "I do not believe that atoms exist."

The metaphysical temptation is doubtless very strong. And Boltzmann inevitably replies to it with the classic counterquestion: If nothing exists behind our perceptions, are we to conclude that a Martian landscape or a planet of Sirius do not really exist because human beings are unable to see them?

And, he adds, "if all these questions are senseless, why can we not dismiss them, or what must we do in order finally to silence them?"

It is up to natural philosophy to clarify these questions, and speaking of the principles of natural philosophy naturally brings to mind the title of Newton's celebrated work. But then, Boltzmann remarks, stating the principles of natural philosophy is equivalent to enunciating theoretical physics. His choosing to speak of natural philosophy, Boltzmann writes, is intended "to show you how little philosophy can cleave to words: these are exactly the same, but today we understand by them something totally different from what Newton did."

A Philosophical Crisis: Mach and Ostwald

As we were briefly retracing certain aspects of Boltzmann's thinking, some readers may have wondered why the traditional themes of a crisis of scientific thought should fail to appear. Their absence is easily explained. According to Boltzmann, there is indeed a crisis, but it is solely a philosophical one. Physics knows no crisis; on the contrary, it is currently delving into "the most puzzling properties deeply affecting our whole view of nature."[21] As a result, scientific *progress* has again placed in question the traditional view according to which physics is divided into a theoretical and an experimental branch. At the root of the problem lies a philosophical error rather than an incurable flaw of scientific inquiry. The old philosophical idea that experimental physics gathers the bricks while theoretical physics builds the house is now crumbling. The tumultuous growth of the physical sciences, Boltzmann writes, has clearly shown that it was a mistake to entrust philosophy with the task of investigating certain problems of a general character, such as the nature of causality and the concepts of matter and force: "Philosophy has been remarkably ineffective in clarifying these questions." Real progress can only be achieved through the cooperation of science and philosophy.

These remarks appear in a lecture that Boltzmann delivered in 1904 at the Congress of St. Louis and was devoted mostly to statistical mechanics. According to Boltzmann, the puzzling properties emerging from the new radiation physics and the fundamental changes needed to understand them are clear indications of the vitality of objective knowledge and of the weakness of philosophy. Now more than ever, he states, it is necessary to formulate new hypotheses that can go beyond "pure facts of observation" and thus lead research to "totally unsuspected discoveries."

To give the scientific revolution even greater freedom, we must fight the thesis that "our theoretical edifices" have been built solely on "logically incontrovertibly grounded truths." The battle against philosophy is for Boltzmann an essential condition for the liberation of physics and therefore his antiphilosophical polemic becomes ferocious: "The most ordinary things are to philosophy a source of insoluble puzzles" because, having taken it upon itself to elaborate concepts like matter, space, time, and causality, philosophy is inevitably led into inconsistent and contradictory statements. "To call this *logic*," Boltzmann remarks with sarcasm, "seems to me as if somebody for the purpose of a mountain hike were to put on a garment with so many long folds that his feet became constantly entangled in them, so that he would fall as soon as he took his first steps in the plains."

The contradictions that the philosophers of matter, space, time, and causality see everywhere, Boltzmann continues, are but "inappropriate and mistaken mental reshapings," the result of elaborating so-called laws of reasoning with uncritical confidence.

The real task given to man's intellect by the new physics is not to "summon [empirical] data to the judgment throne of our laws of thought" but "to adapt our thoughts, ideas and concepts to what is given."

Overestimating those theoretical tools that have shown some validity in certain fields of research, and trying to impose them on new fields of endeavor, has a negative effect on the development of knowledge. Nothing but contradictions and false problems can ensue from this mistaken use of theory, both within theory itself and in its relationship to the world. The external world, in fact, has no obligation to conform to our concepts. "We must not aspire to derive nature from our concepts, but must adapt the latter to the former. We must not think that everything can be arranged according to our categories or that there is such a thing as a most perfect arrangement: it will only ever be variable, merely adapted to current needs. Even the splitting of physics into theoretical and experimental is only a consequence of the twofold division of methods currently being used, and it will not always remain so."

In the lecture of 1904 Boltzmann makes the point that the need to violate the rigidity of certain categories is beautifully illustrated by the changing physical conception of the atom. The triumph of the theory of the electron has shown new and almost unbelievable ways of investigating the nature of the atom that have utterly demolished the notion of its indivisibility. Although the concept of electron is incompatible with that of an indivisible atom, this contradiction has not brought theoretical physics to a standstill. On the contrary, the latter has overcome the contradiction by modifying its concepts and adapting itself to all the new and unexpected facets of nature.

According to Boltzmann, the reason why physics develops without stumbling into the difficulties met by some philosophies hinges on the fact that physical laws are the result of processes of abstraction according to norms rather than of mere observations of facts. Galileo's principle does not derive from the observation of isolated systems since it is impossible to isolate a system from the influence exerted on it by other systems. Similarly, statistical mechanics is an abstract system: its mathematical apparatus frames theorems that follow by deduction, and whose application to nature represents "the very prototype of a physical hypothesis." Being a prototype, such theorems and the theory in which they are contained defy any philosophical scheme since they "boldly" transcend a given experience and present the "facts of experience" in an entirely new light. It is for this reason that genuine theories can stimulate further study and reflection.

As one can see, this is a rather unorthodox and dynamic view of what is meant by mechanical explanation and model. What is left, then, of the ghost that some philosophical histories of science call "Boltzmann's mechanism"? Nothing. The disappearance of this imaginary mechanism —and of the equally imaginary and vulgar metaphysical materialism of which Boltzmann is often said to be an authoritative proponent—enables us to understand the distinction that the great mathematical physicist always makes in his criticism of Ostwald and Mach. Boltzmann does not use the same philosophical arguments against both. In his opinion, Ostwald's philosophy is based on a misunderstanding of Mach's thought. According to Boltzmann, Mach has two basic theses: the first states that we only have correlations between perceptions, analogous to laws; the second states that the concepts of atom, molecule, force, energy, and so on, are simply notions constructed for the purpose of representing such correlations economically. It follows that only perceptions have a predicate of existence, whereas physical concepts are mere additions. However, Boltzmann writes, Ostwald has "understood only half" of Mach's argument. Ostwald has understood only that atoms do not exist, and asks himself: What does? In answering that energy exists and matter does not, Ostwald totally misconstrues Mach's thought, since Mach's contention is that matter and energy are both "symbolic expressions of certain relationships between perceptions and of certain equations among the given phenomena."

In his efforts to produce a philosophy of science Ostwald is prey to confusion. But the fact that his theses are unfounded and superficial does not make them any less dangerous. On the contrary, Ostwald's superficiality is very dangerous because it inevitably achieves a much wider diffusion than the strict and controlled methods of the natural sciences.

Thus the objective point of view is at the very heart of the philosophical battle raging around the sciences. The century-old dispute between materialism and idealism, Boltzmann writes in 1905, cannot be solved by falling back into the inanities of Kant's philosophy or by modern interpretations of "a stupid, ignorant philosophaster" like Schopenhauer.[22] When one has to choose between a materialistic premise to explain the existence of sensations and an idealistic one to derive matter from ideas, it is completely useless to look for verbal refuge in purely philosophical concepts of space and matter. There is only one rational way to confront the problem of space, and that is to frame the concept of space on the basis of appropriate considerations of the axioms of Euclidean and non-Euclidean geometry; and these concepts are not evident a priori. The same is true for the scientists who wish to define the concept of matter. Their problem is not to define this concept philosophically, but to be always willing to modify it in accordance with the developments of physical knowledge.

According to Boltzmann, it is essential to keep a critical rather than dogmatic attitude toward the norms and rules that affect one's willingness to accept change. Against the dogmatism of some atomists and of the sectarian proponents of energetics and phenomenalism an enlightened theory argues that there is no possibility of an absolute explanation. "When I say," Boltzmann writes in 1904, "that mechanical pictures might be able to illuminate such obscurities . . . I do not mean by this that the position and motion of material points in space is something whose simplest elements are completely explicable. On the contrary, to explain the ultimate elements of our cognition is altogether impossible." Nor should one trust implicitly in the mechanical explanation itself. Like every other form of scientific explanation, it too is subject to radical changes. These changes, however, are certainly not going to come about from research in energetics or in phenomenalistic physics: "The ray of hope for a nonmechanical explanation of nature comes not from energetics or phenomenology, but from an atomic [electronic] theory that in its fantastic hypotheses surpasses the old atomic theory."[23]

Flexibility of Boltzmann's Dictionary

Thus far we have attempted to examine some levels of Boltzmann's dictionary. The objective of this study is to identify certain areas of his dictionary and to date some of their developments.

We have sketched a correlation between Boltzmann's thoughts on physics and his physical research, pointing out that throughout the papers of 1871 his approach undergoes a marked change due to the massive infusion of probabilistic formulations into the theory of gases. As a result of the questions raised by probabilistic inferences, as early as 1872 Boltzmann is forced to defend the profound coherence of the application of such inferences to theoretical physics and to argue with Maxwell in support of the nonrenunciatory nature of the new physics.

Thus we observe a series of interesting readjustments taking place in Boltzmann's dictionary. Most obvious is the change in the assumptions he had adopted in his earlier studies: from the thesis that thermodynamics can be reduced to mechanics Boltzmann moves on to the thesis that the calculus of probability is the best approach to understanding the meaning of the second law. In 1872, with the first enunciation of what will later become the H theorem, Boltzmann's theoretical physics vindicates the pre-eminent role of mathematical theory in its most abstract forms.

The date of this first disclosure is quite significant, since general debate on irreversibility and on the apparent contradiction between Boltzmann's physics and mechanics does not begin until after the publication of articles by Kelvin and Loschmidt on the paradoxes arising from the symmetry in dynamic equations. And since Kelvin's attack is published in 1874[24] and Loschmidt's in 1876,[25] it follows that the change in Boltzmann's dictionary, already evident in 1871, is not a result of the general debate on irreversibility.[26]

We do not mean by this that the debate has no influence on Boltzmann's work. What we mean is simply that it affects his work after he has already abandoned the reductionist thesis. It is quite clear therefore that from 1869 to 1872 Boltzmann's dictionary is flexible enough to confront specific problems on its own, independently of any need to defend a research project from the attacks of other scientists.[27]

Such flexibility—together with the relative autonomy we have just mentioned—has implications of considerable importance for the theory of knowledge. The idea that physical theories are essentially open structures[28] rather than logically self-sufficient and rigidly delimited may be quite fruitful when used as a tool for investigating certain phases in the evolution of a dictionary. In the case at hand, although starting from a physics that seeks its justification in a reductionist thesis, Boltzmann's theory is clearly open, in the sense that it can accept modifications dictated by the requirements of mathematics. When Boltzmann questions his own theory and forces it—from 1869–1870 on—to answer in full all the questions that can be reasonably asked with the calculus of probability, the first answer he obtains is that the theory allows for profound structural changes on the basis of the new rules.

This answer may be defined as a turning point, for it implies a stringent criticism of notions and categories that appeared untouchable in the previous versions of the theory.

However, we must be careful with the term turning point, and with all such words meant to describe particular phases in the evolution of a dictionary. Turning point—a term commonly used in certain standard jargons—does not imply a break, or an epistemological fracture, or a fissure in history (to be filled, as some suggest, by injections of psychological cement): turning point denotes a historical process in the full sense of the word.[29]

Around the term turning point, or more or less equivalent terms, revolves the dispute between two contrasting views of history, namely, as a continuous process or as a succession of sudden breaks and logical incompatibilities. Let us consider the turning point in the theory embedded in Boltzmann's dictionary. If we were to compare a monograph written by Boltzmann in 1868 with some of the papers he published around 1876, we would certainly be struck by the substantial difference between them. But if we went on to examine in detail the mass of documents relating to that period, we would be fascinated by the remarkable number of subtle internal changes. And if in the first case one has no right to decree the discontinuity of history, in the second one has no reason to reduce the logic of a dictionary to a monotonic process of small adjustments in the context of continuity.[30]

Whether a dictionary is convulsed by earthquakes or slowly reshaped by minor tremors is a point of little interest to those who wish to understand the dictionary's logic. What is relevant, instead, is the following point: while developing within a vast, unstable, interconnected, and unsynchronized dictionary, a theory is open because some of its fundamental concepts have not yet been rigorously defined.[31] In the case at hand there is indeed a fundamental concept that is still ill defined: the concept of probability.

Although ill defined, the concept of probability by no means represents a weak or erroneous element in Boltzmann's theory. On the contrary, we cannot fail to note that it is precisely the development of the theory that reveals certain unusual or unsuspected properties of probabilistic inferences, and that consequently stimulates a more rigorous study of the concept itself.

It should also be added that in analyzing a physical theory that has reached an advanced stage of completion, one tends to ignore the decades of patient, painstaking work that has prepared the theoretical ground for the final phase of completion. With this observation we do not intend to take exception with the attitude of those who examine well-packaged theories with a certain diffidence toward their history. We only wish to point out that the logic of the past may not coincide with the logic of the finished product without necessarily falling into the fictionalized accounts of either the subconscious or of external events.

The controversy between the partisans of an eclectic version of Newtonianism and the advocates of the mathematical investigation of nature
as a means of overcoming the limitations of the senses is a particular
aspect—weak areas of the dictionary—of a far larger conflict. This conflict
ensues from the need to find explanations for puzzling phenomena and
solutions for unexpected problems, such as arise in the course of research
in electricity and magnetism.

Interesting questions for the historian of physics are posed by the
revolutionary nature of the studies on the classical electromagnetic field
and by the influence these studies have on the scientific view of the world
and on philosophy. For example, how was it possible for scientists still
methodologically bound to the principle of action at a distance to
formulate models and theories that imply action by contact and to
criticize the sophisticated web of conjectures woven by the Laplacian
school about magnetic fluids?

The clash between Kelvin's and Boltzmann's views of mathematics brings
us back to the questions of models and method that were debated at the
end of the eighteenth century and in the first decades of the nineteenth. It
brings us back to Coulomb and Ampère. With regard to models, it does
not seem to make much sense to discuss them in terms of equivalence,
unless we specify that it is a weak equivalence susceptible to well-
formulated variations. The models of Coulomb or Ampère are not

intuitive constructs artificially imposed on physical theories. The models in question imply certain inferences and not others, and these inferences strictly concern theories.

As far as methodological rules are concerned, we find that they are involved more in the justification than in the construction of new concepts. Thus, once we retrace the intricate pattern of Ampère's Newtonianism we still have not made any progress in reconstructing Ampère's electrodynamics or his ideas on the contradictions inherent in the theories of heat and light. It follows that in a dictionary areas of strong rules interact with those weak levels to which physicists like Ampère assign their opinions on what they think they are doing.

The distinction between the autonomy of science and its involvement with philosophy is a problem of the weak and strong interactions that occur between areas of a dictionary, as the dictionary changes in time (that is to say, has a history), in response to the answers given by nature to her questioners and in response to the problems continually raised by theory. In the following we therefore propose the hypothesis that the problem of objectivity is a problem of history.

5   *The Rules of the Good Newtonian*

Ampère and the Theory of Heat

In addition to theories and to the experiments theories suggest, some physicists find it necessary to use models, that is, systems of assumptions mathematically formulated and connected to the theories by particular sets of rules. In a general way, it is pertinent to ask to what extent such models aid scientists in the study of nature. And this question takes on particular relevance when the historian who poses it—to himself or to others—is already convinced that the answer will reveal a fundamental aspect of classical nineteenth-century physics. This conviction stems from a preconceived idea with which we confront that physics. We can summarize this preconception, at least in part, in the following thesis: classical nineteenth-century physics was a sort of mathematicophysical commentary on the principles of mechanistic philosophy. As a corollary, it follows that this commentary on the principles of mechanistic doctrine found expression in the production of mechanical models of the phenomena and objects of the material world. To these models nineteenth-century physicists would have given the signal task of faithfully representing nature. Only one conclusion can be drawn from this thesis and its corollary: classical nineteenth-century physics collapsed because its mechanistic foundations, although presumed to have absolute validity, did not stand up to philosophical criticism. As a result there occurred a great crisis in scientific thought.

It is interesting to note that in this melodramatic view philosophy is granted the traditional privilege of determining what are the objectives, limitations, and virtues of scientific inquiry, and of condemning the latter should it violate certain canons established outside its domain, in what is actually the domain of philosophical speculation. In this manner philosophy succeeds in imposing on science the delusion typical of any doctrine that approaches the dynamics of objective knowledge starting from literary conceptions of man's relationship to nature: the delusion that there are scientific crises caused by philosophical criticism rather than philosophical crises brought about by revolutionary changes in science.

If we dismiss this delusion and proceed to analyze the historical development of the sciences, paying special attention to the topical variations in concepts, in theories, and in the rules for the interpretation of empirical data, what we find is an intricate network of inferences that develop at different levels in response to the often enigmatic answers nature gives to the questions we ask with theories and experiments—inferences in which philosophical categories play a purely instrumental role. A case in point is the question of models in nineteenth-century physics, a question that according to the point of view sketched above should be fully resolved into the problem of mechanism.

According to the so-called mechanistic explanation, Coulomb's or Ampère's theories are a splendid example of the classical dream of representing nature by means of hypotheses concerning natural phenomena, where such hypotheses—or models—are to be based on the eternal laws of motion. However, we cannot accept philosophical speculation on mechanism as the sole proof that this dream really existed. It is true that historical research cannot always rely on documents that speak for themselves and thus solve the problem at hand, and it is true, therefore, that interpretative efforts are often needed; on the other hand, philosophical speculation that turns history into a mere vessel for edifying tales about scientists invented ad hoc is a futile exercise devoid of any real subject matter.

If we ask what was Coulomb's or Ampère's position with regard to models and accept their Newtonian statements without analyzing the actual contents of the writing in which such statements appear, then the answer is plain: Coulomb and Ampère were mechanists. Coulomb formulated the laws of the interactions of electric charges and magnetized bodies with explicit references to the fact that the form of these laws was the same as the form of Newton's law of gravitational interaction. In so doing, Coulomb built a theory based on an explicative model that was purely mechanistic in nature. Ampère, an ardent Newtonian, built electrodynamics on a model of magnetism that reduced elementary phenomena to microscopic interactions at a distance. In case there are any doubts about Ampère's opinions we will recall a passage from his *Essai sur la philosophie des sciences:*

After Newton, mathematicians and astronomers have generally accepted that celestial bodies attract one another at a distance without any kind of contact. All those who are acquainted with the current state of the physical sciences know that the mutual action of the molecules of bodies, including the molecules of the fluid called ether, which is diffused in all of space, occur across the empty, extremely small intervals that separate them. Since it would be impossible to suppose that when two marbles collide the molecules located at the surface of the first are closer to the corresponding molecules at the surface of the second than two molecules of the same marble are to each other, it is evident that the action of collision occurs without contact, owing to the very same repulsive forces that keep the molecules of a body far from one another.[1]

Before we go on to discuss the themes developed in Ampère's theory of magnetism, it should be noted that the first volume of the *Essai sur la philosophie des sciences* appeared in 1834 and that in 1832 and 1835 Ampère published two important essays on physics. In these two essays, and particularly in the second, which elaborates the questions enunciated in the first, Ampère confronts one of physics' fundamental problems, which is closely connected with the question of models.[2]

Having accepted the wave theory of light and the contributions of Young, Arago, and Fresnel, Ampère takes into consideration an objection that, if unresolved, would in his opinion threaten the validity of the physical explanation.

Light and thermal radiation are explained by the wave theory and are governed by identical laws. However, when heat is not transmitted as *"chaleur rayonnante"* but propagates through a body because of differences in temperature between the various parts of that body, the result is not "a vibratory motion that propagates by *waves*" but "a motion that propagates gradually." Hence "an objection is born to the theory of heat propagation by vibratory motions."[3]

Is it possible to eliminate this objection? It is most urgent to do so, since such an objection creates a fundamental problem in physics. Light and thermal radiation can be described by the wave theory, but the law of heat propagation in bodies is the product of the splendid theory Fourier has formulated by rejecting both caloric and wave models as completely irrelevant. The answer attempted by Ampère consists in demonstrating that Fourier's equations can be derived from properly formulated un-dulatory hypotheses. If this can be accomplished, then "one evidently finds the same results" since there is a profound analogy between the various models and Fourier's thesis, which rejects all models and makes heat propagation dependent on differences in temperature. In this case the

objection would no longer cause contradictions in models and its elimination as an objection would re-establish the essential equivalence of the
models themselves, without favoring any one of them.

Once this equivalence is established, will it prevent a better knowledge of
matter at the molecular and atomic level, and will it force physics to study
phenomena solely at the macroscopic level? After reading the 1835 essay
one cannot help ask such a question, particularly in view of the heated
debate taking place in those years on the propriety of formulating
hypotheses concerning the nature of things that are not directly observable.

Now, if we read the 1835 essay from this viewpoint, we find that Ampère's
considerations on the analogy tend first of all to safeguard Fourier's theory
from the objection that would make it incompatible with the wave theory.
"One evidently finds the same results by considering the things we have
just said in both the system of emission and that of vibrations, the quantity
of caloric in the first system having been replaced in the second by the live
force of vibratory molecular motions." The analogy stops here. As far as
Fourier's equations are concerned, it is still true that the wave theory is
"today well demonstrated." Ampère writes, "It is clear that once we
admit that heat phenomena are caused by vibrations it is contradictory to
attribute to heat the repulsive force of the atoms that is necessary for them
to vibrate."[4]

Once the objection is removed, a deeper paradox arises whose solution
requires a careful re-examination of the presumed analogy between
models. The only way to resolve the paradox is to choose the wave theory.
In so doing Ampère relates the wave theory to detailed hypotheses
concerning the structure of matter in particles, molecules, and atoms, thus
establishing a connection between the wave theory and "the principle" of
1814: "In equal volumes of any gas or vapor at the same pressure and
temperature there is the same number of molecules."[5]

Ampère's adoption of the wave theory is a choice that finds justification in
norms and procedures that can be tested in the relationships then existing
between experimental physics, mathematical physics, and chemistry. The
study of these relationships, as carried out by Ampère, is a study that
proceeds according to the rules of physics and chemistry: rules whose
rationality is based on the properties of the theories and on the answers
that nature gives to the questions posed by those theories and to the

experiments planned in accordance with those theories. In other words, Ampère's choice, as stated in the 1835 essay, cannot be reduced to the canons of the Newtonian philosophy that Ampère himself had enunciated in the *Essai* of 1834.

We do not mean by this that Ampère's choice, although dictated by the rules of mathematicophysical theories, is wholly independent from the philosophy of the *Essai*. What we mean to say is that his choice has a degree of autonomy with respect to the philosophy of the *Essai,* in the specific sense that from the latter we cannot deduce the individual steps in Ampère's reasoning; nor could we ever deduce from it Fourier's equations or Young's theses, even though Fourier and Young are both Newtonians. In the dialectic between science's degree of autonomy and involvement with philosophy, the question of whether Ampère is or is not a mechanist becomes simply a badly formulated question. Only philosophy can presume to enunciate interesting answers to senseless questions; but these answers tell us nothing about the actual historical process by which the development of objective knowledge generates concepts and theories that describe the world. It might be argued that the extent of science's autonomy consists only in the following: that deductive acts are mere techniques of reasoning about which all philosophies are in agreement for the simple reason that, as techniques, they do not involve a reflection on knowledge. This approach essentially would reduce science to philosophy by the stratagem of demoting to technique anything that cannot be defined as method. The proper way to write the history of science would then be to write the history of method, leaving the whole body of techniques, including "mathematical techniques," to the footnotes. Even a partial examination of Ampère's dictionary suffices to convince us that, so far as the physical problems of 1835 are concerned, physics cannot be wholly resolved into methodological rules, nor mathematics into technique. The connection between Newtonian method and physical problems is in this case a very weak one.

Violation of the Newtonian Canons

As we were saying, there is a difference between the position Ampère takes in the 1834 *Essai* in support of action at a distance and the manipulations he performs to remove an objection and a contradiction from the physics of his time. The position Ampère takes in the *Essai* defines a general philosophical view and provides a Newtonian language; it is also an attempt to describe the relationship between the individual disciplines and that "natural classification" of science that Ampère considers to be a problem of the greatest import, and to outline the distinction between mathematics and mathematical physics as well as between psychology and ontology. This position, however, is insufficient at any level of its internal structure to prescribe the approach and the rules that should be followed to formulate and then solve a problem of optics in relation to one of thermal radiation.

To remove the objection threatening Fourier's theory and to overcome the contradiction between the wave theory and the caloric hypothesis, Ampère adopts a very specific correlation between the wave theory and the structure of matter.

Ampère's *"particules"* are infinitely small bits of matter consisting of molecules situated at specific distances from one another. The stability of this system is ensured by interactions. The forces involved are repulsions due to the vibrations of the ether, "what is left at that distance of the attractive and repulsive forces of the atoms" forming the single molecules, and Newtonian attraction between molecules. Atoms are material points and their attractive and repulsive forces are extraordinarily intense over small distances.

Thus a distinction must be made between molecular vibrations and atomic vibrations. The former are responsible for sound phenomena; the latter, propagating through the ether, produce heat and light phenomena.

With regard to the problem of vibrations, the atomic-molecular model may be illustrated by using another model that consists of diapasons. In the model of the model we can observe a special case of great interest. Consider a medium containing any number of diapasons *"à l'unisson"* only one of which, or a small group of them, is actually vibrating. The entire system will be progressively hit by waves transmitted by the medium and "the live force of the system of diapasons will be diminishing indefinitely" unless the system is enclosed within a volume bound by a barrier of diapasons kept vibrating "with a constant live force." In the latter variant, the model of the model predicts that the total kinetic energy "will tend to approach indefinitely that of the diapasons enclosed within the barrier without ever reaching it, mathematically speaking."[6]

In the context of the model of the model, mathematically speaking, we may assume that the amount of kinetic energy transmitted between groups of diapasons is proportional to the difference in kinetic energy between such groups. "For the distribution of live force between diapasons," Ampère concludes, "one then necessarily finds the same equations found by Fourier for the distribution of heat . . . starting from the same hypothesis, namely, that the transmitted temperature or heat, which represents here the transmitted live force, is proportional to the difference of the respective values of the temperature."[7] The analysis may be further refined by considering solids vibrating "in three dimensions" rather than simple diapasons. But even in the more complex analysis "all that we have just said still holds true, regardless of the shape or size of the bodies."[8]

It should be noted that in speaking mathematically of models of models Ampère has gone from the physics of action at a distance to that of action by contact, openly violating the Newtonian canons he has himself stated to be valid. The ether that fills the vacuum between atoms and the *"milieu indéfini"* in which the diapasons are immersed are the continuum that enables the "vibrations" to propagate. This continuum cannot conveniently disappear by turning into discrete components, which is precisely what Ampère maintains in the *Essai* when he writes that "the mutual action of the molecules of bodies, including the molecules of the fluid called ether, which is diffused in all of space, occur across the empty, extremely small intervals that separate them." As Ampère fully realizes, transforming the continuum into discrete components for the sake of saving action at a distance only shifts the problem: "To make it easier to understand the analogy between heat propagation in bodies and sound vibrations from solid to solid through the medium of air, I have assumed in this explanation that the molecules of bodies can only transmit their vibratory motions through the medium of ether."[9] However, Ampère adds, molecules can also transmit vibrations directly provided they are so close to one another that the powerful atomic forces will be able to operate. When a molecule starts vibrating and changes shape, the atoms of the nearby molecules, which are sensitive to such a change, will themselves start vibrating. In the realm of metaphors, the canon of action at a distance would thus appear to be safe, despite the fact that Ampère's reasoning has proceeded in the context of action by contact. The author himself admits he has not developed the point of view of action at a distance: "As this point of view requires calculations I have not made, I have not fully developed the consequences of this idea."[10]

To repeat it once more, the content of the 1835 essay is totally incomprehensible on the basis of the criteria expressed in the *Essai sur la philosophie des sciences* of 1834—an effort to which Ampère devoted a great deal of time and intellectual energy over a good many years. Equally incomprehensible, from the point of view of the *Essai,* is the path Ampère followed in criticizing as he did the caloric theory and in accepting the wave theory.

In 1835 the monumental *Thèorie mathèmatique de la chaleur* by S. D. Poisson is published in Paris.[11] In it Laplace's hypothesis of molecules as centers of emission and absorption of thermal radiation is enriched with inferences drawn from the calculus of probability. While Poisson is bringing the idea of caloric to its greatest theoretical lucidity, Ampère undermines the very concept of caloric fluid with the thesis that heat is an undulatory motion. Poisson, a great mathematician, discusses thermal phenomena on the basis of action at a distance. Ampère, whose profound knowledge of mathematics is also beyond question, discusses thermal phenomena, "mathematically speaking," on the basis of action by contact. Although Ampère and Poisson are both good Newtonians, they follow trains of thought that lead to two different physics with respect to heat and light phenomena.

Thus the Newtonians of the Paris school are deeply divided on the question of models. The conflict between Fourier's views and Laplace's, already tense in the early 1820s and further exacerbated by the controversy between Fourier and Poisson, leads to a series of increasingly diverging statements. It is no longer just a question of finding reasons for choosing either the fluid model or the model that equates quantities of heat with quantities of molecular kinetic energy. After the 1780 memoir by Laplace and Lavoisier in which this equivalence is stated,[12] a growing knowledge of physics gradually changes the terms of the problem. The new theories of heat and light as wave phenomena come into play and Poisson's theses are dealt a serious, if limited, blow on a delicate point. Poisson maintains that the wave theory necessarily leads to paradoxes because, if the intensity of light is measured at a certain distance from a small obstacle placed between the light source and the observer, the theory predicts that the measurement cannot distinguish between the intensity observable in the presence of the obstacle and in its absence. But the paradoxical prediction is confirmed empirically and Fresnel's theory triumphs. Poisson's paradox of 1818 disappears from the wave theory, and

Fresnel's wave theory is based on Young's conjectures and on analogous suggestions by Ampère.

In the 1820s and 1830s dogmatic Newtonianism is dealt unprecedented blows. Bessel writes, "Newton's system is not mathematically necessary; its actual existence as the system of nature cannot be demonstrated in an absolute way; it can only be established with a certain degree of probability that depends on the precision of the experiments."[13] The choice between the various models of the nature of heat and light becomes more complex rather than simpler because the criteria one may appeal to for guidance are no longer homogeneous, stemming as they do from theories that are different in both content and scope. In 1823 Avogadro states that the fundamental ideas of both caloric and wave theories "for the time being cannot but be developed in great obscurity and uncertainty."[14] The following year Sadi Carnot remarks in the *Réflexions* that the caloric theory is no longer tenable in view of the new experimental knowledge.[15] In his physics course J. L. Gay-Lussac argues that there are no rational reasons to choose one model over the other,[16] and in an 1829 paper on algebraic analysis Fourier inflicts yet another wound on French physics: "The study of the theory of equations clarifies physical problems."[17] W. R. Hamilton takes a similar position in 1833 when he suggests that the study of mathematical optics should not necessarily be based on Huygen's or Newton's hypotheses.[18] A year later É. Clapeyron asserts that the model-based conjectures of Laplace and Poisson are questionable.[19] This statement clearly reveals the influence of Lamé, who in his physics course at the École Polytechnique adopts Fourier's theses and argues that all models are equivalent.[20]

Ampère's essay of 1835 must be placed in this process full of contradictions rather than in that static caricature some philosophers have termed mechanism, thus reducing the internal labors of classical physics to a series of mere mathematical exercises on Newton's equations and the concept of action at a distance. In an effort to bring this intricate process into a unified picture, in 1830 Herschel seeks certainty in experience, Comte begins publication of the *Cours de philosophie positive*, and Ampère tries to conclude his long studies on the philosophy of science and publishes the first part of the *Essai*. Philosophy has been left behind and the philosophical roots of the mechanistic view of nature have been reduced to slogans among which the new theories creat havoc.

In sum, we are not faced with an unconscious physics that performs mathematical acrobatics with old concepts and attempts to breathe new life into them by means of mechanical models. What we find is a physics that grows strong on disputes and contradictions, on rational debates over paradoxes and sound objections, and on the awareness of the provisional character of models. It is not by chance, then, that Herschel *re*-thinks the concrete bases of the sciences, the mathematician Comte *re*-searches the real structure of scientific knowledge, and the physicist Ampère *re*-views the classical theme of the natural and artificial classification of the disciplines. The common problem is not only to unify the various modes of reasoning but, more important, to realize that the precepts of the Newtonian doctrine have come into conflict with a surprisingly new nature, and to allow the revolutionary power of the sciences to play its part in a new formulation of the philosophical relationship between science and nature. When we say that philosophy is lagging behind, what we mean is that, if philosophy has a task, this task is precisely to understand the general trends in the development of scientific thought, not to prescribe laws for its future development.

The Role of Models in the Foundation of Electrodynamics

Ampère's thoughts on heat and light must be viewed in the context of real physics rather than in the context of mechanistic ghosts. A rereading of the 1835 essay and of the fundamental writings of the 1820–1825 period clarifies the role of models and underlines the correlation between Ampère's theory of magnetism and the ideas enunciated by Coulomb on the subject from 1777 to 1789. The "Ampère problem" may thus be formulated as follows: Why do most physicists accept Ampère's theory but reject the hypothesis that Ampère himself believes to be the foundation of his new electrodynamics? We propose the following answer: in the first half of the nineteenth century the majority of European physicists accept the theory but reject the model because the model does not give the necessary information on the nature of things, whereas the theory provides a useful correlation between phenomena. This answer is valid insofar as those physicists believe that the theoretical correlation between phenomena is a sound deductive correlation, namely, that it is correct, that it is in accordance with well-formulated rules, and that it is confirmed by the available empirical data. In their opinion, its experimental validity is wholly independent of what the model claims to say about phenomena that cannot be measured in the laboratory or observed directly by the senses. This approach is particularly attractive to the physicists of the Laplacian school because, if one has to accept not only Ampère's electrodynamics but his model as well, one is confronted with contradictions in electrostatics and magnetostatics and fundamental doubts are raised over the whole question of potentials according to Poisson. Following this approach, Laplace's school promotes Coulomb's theory to a dogma: it credits Coulomb with having proved the Newtonian nature of the interactions of electric charges and magnetized bodies, and with having demonstrated that these interactions are a necessary consequence of a specific model of electric and magnetic fluids. In reality, Coulomb has never favored any particular model of fluids.[21] After some initial skepticism about Ampère's hypothesis, Faraday will eventually manage to unravel this tangle.[22] His solution will deal a serious blow to the whole French physics of action at a distance by reinterpreting electrodynamics in the context of qualitatively new hypotheses and by the gradual formulation of the concept of field.

Without entering into the merits of the revolution effected by Faraday in the following thirty years, let us return to Ampère and to the role of models in the foundation of electrodynamics. The few pages published in Latin by H. C. Oersted in 1820 pose a general question to mathematical physics that, in essence, is very simple: in the interaction of electric currents and magnetized bodies phenomena are observed that cannot be interpreted as actions at a distance along straight lines, but rather as actions developing in space with circular or vortical patterns.[23] This is what Oersted is trying to say when he speaks of a *"conflictus"* at the basis of all that is observable. This *conflictus* seems to elude Newtonian mathematics, but Oersted himself has been studying it for years, starting in 1813 when he published a monograph on the possible identity of chemical and electric forces.[24] The theory built by Poisson on the notion of potential must therefore be re-examined, notwithstanding the fact that from 1812 to 1813 it has found an apparently sound basis in Laplace's equations and in his theorems on the attraction between spheroids, and that it has been successful in arriving by deduction at some of Coulomb's fundamental results. Poisson's masterpiece continually refers to a discrete structure of electric and magnetic fluids, where the elements of such fluids interact with forces at a distance instantaneously established along straight lines. The problem is that his theory cannot account for the circular motions of Oersted's *conflictus*.

The solution of the problem, formulated by Ampère after months of intense work, develops along several lines of attack, and without ever abandoning the Newtonian manifesto, yet succeeds in demolishing Poisson's theory. At first sight Ampère's writings appear to abide by two well-known criteria: electrodynamics is a body of propositions that, on the one hand, have an empirical content rigorously verifiable in the laboratory, and, on the other, are linked by deductive chains whose validity is ensured by the rules of calculus. How could there be any doubt about the validity of a deductive chain that is both correct and endowed with empirical content? Or about Ampère's Newtonian faith, when in his most comprehensive paper published in those years he states, "Guided by the principles of Newtonian philosophy, I have reduced the phenomenon observed by Oersted to forces always acting along the line that joins the two particles between which such forces are exerted."[25] Yet, Poisson's theory is also a system of correct deductive chains endowed with empirical content. It, too, is guided by the principles of Newtonian philosophy, and it, too, involves actions at a distance. Is it possible, then, that the basic difference between Ampère's and Poisson's deductive chains is that they make different predictions in the area of phenomena inaugurated by Oersted?

A comparison of the two theories necessarily leads to a negative answer to this question. The two theories are radically different from each other, and not only in their ability to make predictions: these theories actually give physicists two distinct versions of what happens in nature. The choice given to physicists is very complex because it is not just a matter of deciding between two mathematical tools that make different predictions about certain empirical facts. The choice implies two different views of the physical world, that is, of that body of statements with which physicists describe the phenomena. It is not by chance that Laplace refuses to accept Ampère's hypothesis: his refusal is dictated by the fact that such a hypothesis turns upside down his "way of looking at things."[26]

Two basic decisions have to be made: whether the physics of action at a distance is powerful enough to resolve Oersted's *conflictus*, and whether there really exists in nature such a thing as a magnetic fluid. The second horn of the dilemma is just as relevant as deciding whether a planet, a star, or a tree really exists.

Let us consider the title of the celebrated essay published by Ampère in 1827: "Mémoire sur la théorie mathématique des phénomènes électrodynamiques uniquement déduite de l'expérience."[27] What exactly does it mean "to deduce a mathematical theory solely from experience"? And how does Ampère interpret Newton's fourth rule, which states, "In experimental philosophy *propositions derived by induction from the phenomena,* despite contrary hypotheses, are to be considered true, either rigorously or as far as possible, until new phenomena appear by which they are rendered either more rigorous or susceptible to exceptions." And Newton's statement in the "Scholium" that "in this philosophy *propositions are deduced from the phenomena* and generalized by induction"?

It is simply not enough to accuse Oersted, more or less explicitly, of being a neo-Cartesian. Has Oersted derived mistaken propositions from the observed phenomena? In what sense are they mistaken? Because they are deduced from the phenomena, obtained by induction from the phenomena, or inferred demagogically from the blunders of *Naturphilosophie?*

According to the first paragraph of Ampère's "Mémoire," the *conflictus* can be resolved if one knows "the approach to follow in the investigation of the laws of natural phenomena and of the forces that produce them." Such an approach gives precise indications as to what should and should not be done. One should not think that a law may "be invented starting from more or less probable abstract considerations." Thus no hypothesis or mechanical model should be devised for the purpose of constructing a theory on it. This is what should not be done. Turning to what should be done, Ampère writes, "First observe the facts and change the circumstances as much as possible; couple this initial work with precise measurements so as to establish general laws based solely on experience; then deduce from the laws thus obtained, independently of any hypothesis concerning the nature of the forces producing the phenomena, the mathematical value of these forces, that is, the formula that represents them. This is the way followed by Newton."[28]

This is neither the fourth rule—You shall derive propositions from the phenomena by induction—nor the norm from the "Scholium"—You shall deduce propositions from the phenomena and then generalize them by induction. Ampère's rule is, instead: You shall derive general laws from the phenomena by induction and then deduce the mathematical formulas. Ampère writes, "To establish the laws of these phenomena I have consulted experience alone, and I have deduced from it the only formula representing the forces to which they are due." And then he adds, "I have not done any research into the causes to which such forces may be attributed, being persuaded that any research of this type must be preceded by a purely experimental knowledge of the laws and by the determination, deduced solely from these laws, of the value of the elementary forces, whose direction is necessarily that of the line joining the material points between which the forces are exerted."[29] But these are justifications that do not correspond to the road actually followed by Ampère. They are really considerations expressed in deference to methodological rules, since it is hard to believe, in any case, that experiments speak for themselves to the observer who confronts them without a preconceived theory. While in no way explaining the genesis of electrodynamics, they remain a justification in the full sense of the word: they tell us that Ampère considers himself a loyal Newtonian and intends to convince everybody else of his orthodoxy.

When Ampère proclaims the independence of electrodynamics from models he is very specific about it: this independence concerns both the models that can be made before framing the theory and the models that can be made after its completion, and applies to every well-formulated theory. The body of Kepler's laws and Fourier's theory, Ampère writes, rest solely "on general facts taken directly from observation" and furnish equations that are "an exact representation of the facts." Thus, "The main advantage of the formulas obtained in an immediate way from general facts, and determined by a number of observations large enough so that their certainty cannot be questioned, is that such formulas remain independent of both the hypotheses that may have aided their authors in the search for such formulas and of the hypotheses that might replace them later on."[30]

But if all this is true, what is the basis for Ampère's statement that "magnets owe their properties to electric currents circulating around each of their particles"? It is hard to believe that this is a general fact derived immediately from observation, or a postulate inferred from the observation of a sufficient number of events. Yet Ampère is profoundly convinced of the physical truth of his hypothesis. In a letter to Roux dated February 1821 he writes that the difficulties that arise from postulating the "identity of electric and magnetic fluids" and the "existence of electric currents in magnets" are due to prejudice.[31] At the root of this prejudice lies Coulomb's idea that there cannot be any interaction between electricity and "the presumed magnetic fluids." This idea, Ampère remarks, has been accepted "as a statement of fact."

Ampère's judgment is too severe if for no other reason than that he owes much to Coulomb. But his opinion of the other members of the Institut is quite instructive: "It is really strange to see the efforts some intellects make to reconcile new facts with the gratuitous hypothesis of two magnetic fluids different from electric fluids, simply because they have not yet adjusted their ideas to them!"

The structure of the theory Ampère is constructing in his great essay on electrodynamics emerges clearly after his declarations of Newtonian orthodoxy. The theory is fairly general at the mathematical level, and starts not so much from facts as from the analysis of interactions that Ampère assumes to be "inversely proportional to the power n" of the distance. After long chains of deduction, the author arrives at the central question of the identity of solenoids and magnets—a question of demonstrations. But it is here that conjectures are relevant. If we accept the theory of the two magnetic fluids, Ampère notes, we have to discuss four forces interacting between pairs of "magnetic elements" and two forces interacting between a magnetic element and an infinitely small piece of conducting wire. However, if we substitute each pair of magnetic elements with a solenoid—which amounts to formulating a bold hypothesis concerning the microstructure of matter—all the phenomena can be resolved into one principle. Although the calculations may arrive at identical results in both cases, "it is not in these calculations or in these explanations that one should look for either objections to, or confirmations for, my theory."[32] The distinguishing factor, according to Ampère, is given by the reduction to a single principle of all three fundamental macroscopic interactions: between magnets, between currents, and between magnets and currents. That this distinguishing factor is not introduced only for simplicity's sake is demonstrated by another general characteristic of Ampère's theory, discovered in 1823 by Felix Savary: whereas Ampère's law cannot be deduced from Coulomb's law, the latter can be deduced from the former.[33]

The validity of Ampère's hypothesis and the relationship between his hypothesis and the mathematical structure of the theory cannot be decided by a *querelle* over the merit of Newtonian method, but rather by mathematical demonstrations capable of going beyond the apparent equivalence of the various models and of showing that the new formulas, together with Ampère's model, lead to a deeper understanding of matter.

This deeper understanding does not lead to elementary constituents, however. Ampère is quite clear in this respect, and in his statements we already see a link with the thoughts that will lead to the 1835 essay on heat and light. If the simplest force that appears in the new electrodynamics is the force between elements of current, Ampère writes, and if the other forces (between currents and magnets and between magnets) are "more or less complex derivatives" of the first one, "can we conclude that the first is to be considered truly elementary?"

The answer is unequivocal: "I have always been far from thinking so, and in the *Notes sur l'exposé sommaire des nouvelles expériences électromagnétiques*, published in 1822, I tried to explain it to myself with the reaction of a fluid diffused in space whose vibrations produce light phenomena; I said that it should be considered 'elementary' in the same sense as chemists place in the class of simple bodies all those they have not yet been able to decompose, even when it could be assumed, because of some sound analogy, that they are in fact compounds."[34]

Ampère's model never professes to reflect passively the ultimate elements of matter, and this is a fundamental trait of the physics outlined in the "Mémoire." Going back to the statements of Newtonian orthodoxy that precede the presentation of the deductive work, some additional remarks are now in order. How are we to understand Ampère's persistent appeal to observation and how does it fit into the theoretical framework?

Physics and Philosophical Rules: Aepinus, Coulomb, and Ampère

We should recall that the Laplacian approach to research differs substantially from the indications given by eighteenth-century students of rational mechanics, and that the influence exerted by Laplace on French mathematical physics in the first decades of the nineteeth century is very strong even among the scientists who reject significant portions of Laplace's method. Laplace maintains that the process of mathematization of mechanics should receive increasing support from observations and experiments. To render scientific knowledge more secure and to ensure its continual progress, Laplace argues, it is necessary to obtain ever more accurate laboratory measurements.

This does not mean that Laplace questions the validity of the approach outlined by Lagrange in the *Mécanique analytique* of 1788, which consists of "reducing the theory of this science, and the art of solving the problems relative to it, to general formulas, whose simple development furnishes all the equations necessary for the solution of each problem." At the same time, however, Laplace maintains that the science of heat, for one, cannot advance without good calorimeters. Hence his collaboration with Lavoisier and the letter of 1783 to Lagrange in which he justifies himself for having gotten the notion of doing experimental research.[35]

Laplace's attitude is symptomatic. With him and after him, French mathematical physics meets and clashes with the notion of systematic experimentation and searches for a way to join theory and practice in accordance with Newtonian precepts. In the process the French physicists of the École are confronted with a new situation in physics and attempt to find rational variants for the body of physical and mathematical notions they have at their disposal. This process of revision of Newtonian philosophy is set in motion by problems inherent in the natural sciences. As early as 1777, for instance, a monograph entitled *Nouvelles expériences sur la résistance des fluids* is published in Paris by the Académie des Sciences at the request of Turgot. This monograph is the work of a committee that includes such scientists as d'Alembert, Condorcet, Bossut, Legendre, and Monge. It is very significant that the monograph should state that "the art of interrogating nature through experience is very delicate" and that "there is no science without reasoning, that is, without theory."

It is in the context of this particular relationship between mathematical theory and the art of interrogating nature that C. A. Coulomb, an engineer and a physicist, produces his great essays. Coulomb is also a staunch Newtonian. In the "general corollary" to the "Recherches sur la meilleure manière de fabriquer les aiguilles aimantées" (1777) he states, "It thus appears confirmed by experience that it is not vortices that cause the different magnetic phenomena, and that to explain them we must necessarily invoke attractive and repulsive forces of the same nature as those we are obliged to use to explain the weight of bodies and celestial physics."[36] To give a concrete explanation of this fact it is necessary to experiment in very special conditions, and to build the devices suited to this task, it is in turn necessary to develop some sections of theoretical physics. Consequently, the "Recherches" of 1777 are followed in 1784 by the "Recherches théoriques et expérimentales sur la force de torsion et sur l'élasticité des fils de métal"[37] and from 1785 to 1789 by the memorable series of *Mémoires sur l'électricité et le magnétisme*. The first of these *Mémoires* is a consequence of the "Recherches" of 1784. It concerns the construction and use of "an electric scale," which is the starting point for a new formulation of the theory of electricity and magnetism since it enables Coulomb to confirm a proposition he now enunciates as the fundamental law of electricity: "The repulsive force of two small spheres electrified with the same kind of electricity is inversely proportional to the square of the distance between the centers of the two spheres."[38]

Now, what is Coulomb's position with regard to the models that describe the nature of electricity and magnetism? Once the hypotheses of vortices have been eliminated as contrary to the principles of mechanics, does Coulomb feel that he has to choose a Newtonian model that will explain the interaction of electric charges and magnetized bodies?

On the basis of the texts we have to conclude with Stewart Gillmor that Coulomb in no way regards this problem as fundamental.[39] In the fortieth paragraph of the sixth *Mémoire,* published in 1788, Coulomb writes that two models are currently available.[40] The first envisages two electric fluids whose parts interact by means of Newtonian forces. The second, proposed by Aepinus, envisages only one fluid, which interacts with matter by means of the same forces discussed in the first hypothesis. Although the first model is preferable, there still remains the fact that "as far as calculus is concerned, the supposition of Mr. Aepinus gives the same results as the hypothesis of the two fluids." Consequently, instead of making a definitive choice between the two models, it is more expedient to analyze their degree of probability. Coulomb comes to the following conclusion: "Since both explanations have but a more or less high degree of probability, I should explain, in order to protect the theory from all systematic disputes, that in the hypothesis of two electric fluids my only intention is to present the results of calculations and experiments with the least possible number of elements, and not to indicate the true causes of electricity."[41]

This theme reappears in the seventh *Mémoire* (1789),[42] in which Coulomb writes that "every hypothesis of attraction and repulsion according to any law whatever should not be considered other than a formula expressing a result of experience."[43] Even though a given hypothesis of molecular interaction may permit one to deduce theoretical results with the full support of measurements, we cannot hope to attain a deeper knowledge of the causes unless we build a larger theoretical structure—"a more general law"—capable of gathering into one deductive framework physical properties that currently appear totally unrelated. Once again, it is not a matter of choosing between two models. Although both models lead to identical conclusions confirmed by facts in most of the observable phenomena, still "we must confess" that there are cases that elude either model and in which "theory is found to be in contradiction with experience."[44] Since both models are weak, the problem is therefore to modify both of them.

The *simplest* modification is "to assume that in Aepinus's system the magnetic fluid is forced to remain inside each molecule," still assuming, however, that the fluid is able to move and be carried from one end of the molecule to the other. In this case each molecule has two poles and behaves like an infinitely small magnet.

Coulomb's variant of Aepinus's model plays an extremely important role in the physics of the first two decades of the nineteenth century. Many inferences essential to Poisson's mathematical work are based on Coulomb's model, and so are the additional variants elaborated by J. B. Biot and presented in his *Précis élémentaire de physique*.[45] The idea of molecular polarization implies that the microstructure of matter is constituted as a complex of objects, or molecules, which behave like magnets. It is from this point of view that Coulomb modifies Aepinus's one-fluid model and that Biot and Savart later attempt to explain the multiplicity of electrodynamic phenomena by ascribing them to micro-magnets lined up along conductors traversed by electric current.

In a certain sense, Ampère also builds a variant of Coulomb's model: he accepts the idea that molecules behave like magnets but develops a different explanation for *how* they behave. Ampère modifies Coulomb's variant of Aepinus's model by building another model that eliminates all magnetic fluids from the body of objects present in nature and explained by physics. The most significant aspect of Ampère's version resides in the identification of a limiting factor in Coulomb's variant of Aepinus's model: the existence of two distinct fluids (electric and magnetic) necessarily implies a lack of interaction. In the letter of February 1821 to Roux Ampère writes, "You are quite right to be surprised that people did not experiment twenty years ago on the action of the voltaic pile on magnets. The reason for this is that Coulomb's hypothesis concerning the nature of magnetic action was taken as an accepted fact. It absolutely rejected the idea of any action between electricity and the hypothetical magnetic fluids."

It is evident that neither we nor Ampère could base this laborious process of critical re-elaboration of past knowledge and of construction of new concepts—the birth of electrodynamics—simply on the correct application of Newtonian canons (or, rather, on the application of a Newtonianism that in Ampère turns into the spiritualism of Maine de Biran and the "eclectic philosophy" of Victor Cousin). The canons whose virtues Ampère celebrates in the introduction to his work on electrodynamics have in fact different roots: on the one hand, they draw nourishment from the illuministic tradition, and, on the other, they absorb the philosophical food that Cousin distills first from the pages of Thomas Reid's *Common Sense Philosophy* and the spiritualistic theses of Maine de Biran and then from the speculations of Hegel and Schelling. It is these anti-illuministic roots that turn Ampère against Laplace and explain the difference that exists between the Newtonianism of the father of electrodynamics and the Newtonianism of the author of the *Traité de mécanique céleste*. The former regards metaphysics and the question of god as a very important and autonomous area in which to exercise the tools of reason; as he writes in the *Essai*, "What indeed would a treatise on philosophy be if it did not speak of God?" The latter, instead, views science and its progress as the means of gradually reducing what is left of metaphysics, myths, and superstitions. The former seeks a way to reconcile science and religion, the latter fights for the victory of lay reason.

Such differences and contrapositions in the philosophies of scientists become understandable in the historical reconstruction if we keep in mind that scientists often oscillate between the various positions suggested to them by the problems and solutions of a growing objective knowledge, and between the systematic indications they seek in one philosophical doctrine or another. In this manner scientists help to modify, sometimes in a radical way, both cultural trends and the view of man's relationship to nature that entire social strata have—or wish to have. Philosophy lives solely on these forms of mediation, which enable it to appear in the guise of a real methodology and, as such, to influence certain scientific practices. And this is all there is to say about the interaction between scientific thought and philosophical speculation.

Starting from this interaction, however, we cannot predetermine the actual scientific steps leading from Coulomb's model to its variants according to Poisson and Biot and eventually to Ampère's reframing of the theory. Being aware of science's involvement with philosophy does not entitle us to reduce the rational development of models, theories, and measurements to a philosophically determined process, to acts of obedience to methodological canons, to mechanistic precepts, or, worse, to Newtonian rules. In the course of scientific inquiry reason meets and clashes not only with what is enunciated by method but with what exists in the world. Thus in each new situation a different network of interactions forms between involvement and autonomy, and the various phases in the growth of objective knowledge have to be analyzed separately, that is, topically. Historical research, as a rational reconstruction of the local webs of interactions between science and philosophy, has real problems to investigate, but it will encounter them only by destroying the philosophical metaphors that turn such problems into methodological labels. In short, when we say that Ampère was a Newtonian and Oersted a romantic we have said nothing about the origin and content of electrodynamics.

The Problem of Electricity from Nollet to Cavendish and Coulomb

In retracing some of the developments of scientific thought we have
arrived at the end of the eighteenth century. If we now go back a little
further in the past we find a very different physics: a physics that around
the middle of the eighteenth century seeks qualitative answers by manipu-
lating hypotheses and inferences over an ocean of observations and
empirical data; a physics that, contrary to what has been happening for
decades in astronomy, rational mechanics, and optics, is not yet capable
of investigating nature with mathematics and of "pruning down" the
complexity of the phenomena in the Galilean manner.

Only a few years separate Nollet from Cavendish, Aepinus, and Coulomb;
yet in those few years the empirical ocean is radically reduced, subdivided
and reshaped by means of systematic deductions that the new theorists
borrow from mathematics and from that theory of motion many regard as
nothing more than a branch of analysis.

During the last decades of the eighteenth century the formalized theories
of electric and magnetic phenomena constitute the triumph of the tenet
that all is motion and, at the same time, the beginning of a stringent
critique of the notion that nature actually obeys such a tenet. Re-
examined with the new tools, Nollet's empirical ocean appears to
theoretical physicists as a specialized problem that can no longer be fitted
into any one model.

6   *The Empirical Ocean and the Conjectures of the Abbé Nollet*

Common Sense, Logic, and Mathematics

The so-called equations of Newton were written by Leonhard Euler around the middle of the eighteenth century, sixty years after the first edition of Newton's *Principia*.[1]

The fact that it took so many decades to arrive at the actual writing of Newton's equations is not the kind of thing that a strict "internist" can easily relegate to the footnotes, having described in the text the methodological development of a research program based on a particularly well-established foundation. On the other hand, the position of the strict "externist" is no more tenable in this case. In the first place, he would have to decide whether those decades were a period of normal science or a time of profound changes due to Euler's genial flashes and psychological crises. In the first case, Newton's equations would be the result of plain routine work, somewhat in the nature of solving a crossword puzzle; in the second, they would appear as some sort of magical flowers that suddenly burst into bloom in the recesses of someone's mind. Lastly, it would be even more startling to learn the economic reasons why it became imperative to write on a piece of paper that $F_x = ma_x$.

The complicated dictionaries of scientists like Johann and Daniel Bernoulli, Alexis Claude Clairault, Jean Baptiste le Rond d'Alembert, and Leonhard Euler are at work in those decades. Each of these dictionaries consists of a remarkable number of papers, letters, books, and discussions about the most diverse subjects, including, to mention just a few, theories of vibrating ropes, attempts to solve particular differential equations, conjectures on hydrodynamics, hypotheses concerning the structure of matter, reflections on Leibniz's calculus and on the calculus of fluxions, as well as a veritable sea of considerations on the meaning of science. With such rich dictionaries, it would be hard to maintain that the process of mathematization is a translation in the sense suggested by Poisson in 1835, that is, a deductive process that adds nothing to, and subtracts nothing from, the subject to be translated.[2]

On the other hand, it would not be more helpful to think of mathematization in terms of "unfaithful translation." Mathematics would do our knowledge of the world a terrible turn if it were to translate it into formulas unfaithfully. Moreover, if this systematic infidelity had been systematically confirmed, we would have been aware of it for centuries.

There is a variation of the concept of translation that would appear to solve our problem. This variation was suggested by Pierre Duhem in his critique of Maxwell's physics.[3]

According to Duhem, physics advances by a "double movement" in which "common sense and mathematical logic operate together and inextricably mix their own particular procedures." From this point of view, common sense reigns supreme in the realm of empirical laws, where it separates true from false, while mathematical deduction governs the field of schematic representation. There is yet a third region, separating the realm of common sense from the realm of mathematics, in which "the communication between observation and theory is assured." It is in this region that physics grows.

Once this is stated, however, Duhem's variant has not yet contributed anything. Duhem goes on to explain that a genuine physical theory is a soundly formulated theory and that a soundly formulated theory is one in which "algebraic calculus" plays a purely "auxiliary role." This means that the set of deductive moves of mathematics applied to physical theory may at any time be replaced "by the purely logical reasoning of which it is the shorthand expression."

These guidelines are supposed to prevent theoretical physicists from betraying reason, and what is meant by "betraying" is the artificial substitution of logic with calculus. In Duhem's opinion this is the basic flaw of Maxwell's physics. When Maxwell mathematizes theories he introduces "algebraic models" into them, and in this sense makes the same mistake as Kelvin, who uses "mechanical models." The same erroneous artifices that appear in mechanical models are in fact employed in algebraic models, with the result that knowledge falls to the level of fantasy.

Duhem's variant has a particular characteristic: mathematics applied to physics is presented as an auxiliary to logic. Are we then to understand that the formal treatment of a theory involves its logical structure rather than the mathematical translation we started from? To put it another way: In shifting the problem of framing a theory from mathematics to logic, does Duhem's variant clarify the relationship between theory and experience?

Duhem's answer is quite candid: "Without aspiring to explain the reality hidden beneath the phenomena whose laws we are grouping together, we perceive that the groups established by our theory correspond to real affinities between the things themselves."[4] Duhem then adds that the more perfect a theory becomes, the better we perceive that the logical order in which it organizes experimental laws is the reflection of an ontological order, although we cannot prove that it is so.

We are now back exactly where we started, that is, at the point where we asked ourselves why the mathematization of a theory should be such a lengthy and laborious process rather than a routine translation into symbols of what we are supposed to have already learned from the senses and from experiments.

We would have the beginning of a solution to this type of problem if we could show that a process of mathematization is neither a translation, faithful or unfaithful, nor a purely formal substitution of the logic pre-existing in empirical laws, in other words, if we could show that mathematics is neither a different language from the one we use in enunciating facts, nor an auxiliary tool with respect to a deeper logical structure in the statements on facts. During the process of mathematization, the logical structure established between descriptive statements about facts by means of various inferences is subject to sometimes radical modifications that not only change the form of the theory but affect the empirical evidence itself, and thus force us to give a different interpretation to the observations and to the relationships between measurements. If this is true, then there is all the more reason to say that there are no translations but, rather, reformulations of laws and reinterpretations of facts, as well as changes within the very field of applicability of the nonmathematized theory.

The revolutions that mathematics brings about in physics are particularly interesting when we are dealing with the development of theories and concepts entailing a massive intervention by mathematical strategies in a relatively short period of time. A case in point is the research in electricity and magnetism in the eighteenth century. Only a few years elapse from the *Essai sur l'électricité des corps* of Jean Antoine Nollet (1746) to Coulomb's

"Recherches", and yet during those years the theories undergo changes of fantastic proportions.

It is in the context of this particular research that we will now examine some aspects of the problem we introduced earlier with our remarks on the so-called equations of Newton.

Precaria, & ex Hypothesi Petita

"If I try to guess what I cannot see, I want my conjectures to be based on what I have seen."[5] These words of Nollet refer to the attempts he makes around the middle of the eighteenth century to prevent his investigations in electricity from becoming mere descriptions of phenomena and to understand nature's own "mechanism." Nollet does not confine himself to enunciating his many experiments. He states that to perform experiments without falling into error it is essential to have a theory, "whatever those savants may say who give it as their opinion that no theory should be entertained until all the facts have been exhausted and not a shade of contradiction remains amidst them."[6]

According to Nollet, a scientist's first task is to eliminate all false theories and to formulate a true one. Theories based on vortical hypotheses are false, while theories based on the notion of attraction and repulsion are true. The criterion for distinguishing truth from falsehood consists of making recourse to unquestionable phenomena and to the principles of physics. In Nollet we thus see the total defeat of the Cartesian approach at the hands of the Newtonian dogma.

Turning now from the enunciation of the dogma and its virtues to practical application, we find that the network of nonmathematical inferences in Nollet's dictionary is surprisingly rich with internal correlations, however crude his theoretical structure may be as compared to those developed a few years later by scientists like Aepinus, Cavendish, and Coulomb. What we mean is that Nollet's theory selects and organizes data according to correlations of very great scope, although they are not enunciated in the more rigorous form that characterizes Coulomb's approach, for instance. Nollet erects inferences upon an ocean of experimental data and finds experimentally verifiable connections between widely separated groups of facts in this ocean of available data. This network of correlations conjoins all that Nollet knows about electricity, heat, light, chemistry, the growth of vegetables, human physiology, hydrodynamics, and the treatments for paralysis.

The reason why it is interesting to reconstruct at least part of Nollet's physics is that Coulomb and Ampère built their physics on that of scientists like Nollet, and although they obtained marked successes they drastically reduced its scope by shifting its theoretical objectives and by removing as mathematically false specific laws and hypotheses that Nollet regarded as rational and empirically sound.

A particularly interesting text in this respect is the *Recherches sur les causes particulières des phénomènes électriques, et sur les effets nuisibles ou avantageux qu'on peut en attendre.* This essay in five "discourses" was written by Nollet in 1749 to defend the *Essai sur l'électricité* from criticism and to state the rules for the understanding of electric phenomena in general.[7]

In defending his theory Nollet discusses the question of explanation. "I have particularly applied myself to the consideration of whether this theory succeeds not only in accounting for the principal phenomena . . . but in explaining their circumstances and the effects they produce; I am convinced that if the mechanism of electricity is truly the one I have envisaged, this first key will little by little put me in possession of others and lead me to a deeper insight into Nature's secret."[8]

To begin, let us consider the arguments Nollet advances in support of the hypothesis that electricity consists "of the contrary and simultaneous motions of two substances, outflowing and inflowing," a hypothesis that leads him to view "the state of a rubbed or electrified body, from which electric emanations proceed, as a condition or, if one prefers, as the immediate cause that produces the two motions."[9] The objection Nollet wishes to refute asserts that the inflowing matter is "a purely hypothetical substance": *precaria, & ex hypothesi petita.*[10] Now then, Nollet asks, is a hypothesis necessarily gratuitous? When we postulate, on the basis of repeated observations, that there is something that returns to electrified bodies, that this something is an electric matter originating from nearby bodies and not the original outflowing matter that the air surrounding the body pushes back onto it, we certainly leave ourselves open to the possibility of objections.[11] But if these objections earnestly aim to demonstrate that the inflowing matter is *precaria, & ex hypothesi petita,* they must satisfy three specific conditions. First they must prove that the hypothesis is "useless"; then that it is "not known except as a supposition"; finally "that there is a manifest contradiction in having an electric matter issuing from bodies that are not electrified."

Of the various experiments and common observations that refute the objection by demonstrating that in no way does it satisfy the three conditions, Nollet mentions those relative to the observable effects of electricity on fluids. The rate of evaporation and outflow in fluids descending through capillary tubes shows values higher than normal when the fluids are electrified. This phenomenon is commonly attributed to the action of the outflowing matter. But unless we assume an inflowing matter, how are we to explain the fact that "these same accelerations are observed, as I have done and as everybody can do, in bodies that are not electrified but are situated near others that are"?[12]

The results of these experiments are corroborated by observations performed on organic matter, and in particular on man. "If I cause myself to be strongly electrified," Nollet writes, "and if a nonelectrified person holds up his fingers or a sword a few inches away from me, whence I observe a luminous flux to come or I feel a strong wind to blow, am I still making a gratuitous assumption when I say that matter originating from these bodies is flowing onto me?"[13]

According to Nollet, it can only be concluded that the objection is groundless, since it can demonstrate neither that his conjecture is based on a mere hypothesis and therefore useless, nor that the assumption of an outflowing matter is by itself capable of explaining all the observed phenomena. In order to make the assumption of an inflowing matter completely useless, we would have to postulate that the outflowing matter is sent back onto electrified bodies by the surrounding air. In this case, however, we can no longer account for the fact that nonelectrified bodies are still capable, under certain conditions, of emitting "continuous jets of a fiery matter altogether similar in color, smell, etc., to those issuing from a rod of electrified iron."[14]

Thus the objection violates empirical evidence. Moreover, it is based on the assumption that the air acts upon the outflowing matter. And since this assumption in turn implies the notion of *"vortex aëreus,"* the objection as a whole violates both empirical evidence and the laws of physics. Consequently, any test based on "the principles of physics and on phenomena that are beyond doubt" will prove it utterly wrong.[15]

It is certainly right to consider Nollet a strong supporter of the view that "in electricity, as in every other branch of physics, it is on the evidence of our senses that we judge things."[16] But it would be wrong to overlook the fact that for Nollet a judgment on things, although based on the evidence of the senses, must be guided by rational arguments, that is, by a theory The second discourse is devoted to the "rules to be followed in order to determine whether a body is electric, or whether it is so in greater or smaller measure." In our experience, Nollet remarks, we find different manifestations of electricity and we should consider them carefully. By itself, however, the observation of these manifestations does not remove the prime cause of errors. Observation will lead to a correct judgment only *"if we conceive of electricity according to a certain idea,"* whereas the wrong judgment will ensue *"if we conceive of electricity as a virtue that resides in electric bodies."*[17]

Although the *Recherches* are apparently nothing more than an exposition of qualitatively discussed experiments, what we learn from them is that the correlation of facts is determined by theory. In effect, Nollet gives the reader a particular theoretical interpretation of experiments that could be planned, performed, and studied in a different way by proponents of different theories. Far from suggesting that experience is devoid of any genuine value because the questionable product of conflicting theories, the fact that empirical evidence can be modified in the light of theory tells us that a new objective problem is emerging from the classification of bodies into electrified and nonelectrified. In the light of Nollet's theory the existence of nonelectrified bodies is a very serious problem since there are observable interactions between electrified bodies and nearby bodies that have not been electrified. Thus the theory must be framed in such a way as to allow for a rational formulation of this fundamental problem and its solution. Newtonian method is of little help in this respect, in the sense that a rational distinction between electrified bodies (by rubbing or by direct contact) and nonelectrified bodies cannot be deduced from it. This forces Nollet to draw inferences from a veritable ocean of observations on objects and animals. Here again, methodology in itself does not say which observations are relevant and, on the other hand, observations do not carry labels with their meaning printed on them. This point will become clearer as we follow one particular development of Nollet's theory, which, starting from a hypothesis on electric matter, arrives at a possible treatment for paralytics. It would be wrong to consider this part of Nollet's theory as some sort of curiosity or as a misguided attempt to discuss irrelevant facts in physical terms. All of these problems are of the greatest importance to a physicist working about mid-eighteenth century and persuaded, as we learn from the *Recherches,* that "there are no established rules in physics that a decisive experiment cannot abolish or curtail."[18]

The Net of Inferences

In the fourth discourse Nollet remarks that although current knowledge of electricity does not yet permit the use of "geometrical expressions" to determine, for instance, "whether electricity flows in proportion to the mass or in proportion to the surface,"[19] this consideration should in no way hinder research on this problem. In the fifth discourse Nollet examines "the effects of the electric virtue on organized bodies."[20]

How are *"corps organisés,"* that is, vegetables and animals, made? Nollet's answer is fully in accordance with mechanistic philosophy. In 1689 Malpighi had studied "the machines in our bodies" and regarded organisms as "consisting of ropes, filaments, beams, running fluids, tanks, canals, filters, sieves, and similar machines." Sixty years later Nollet examines "organized bodies, which in some respect may be considered hydraulic machines built by nature itself."[21]

Now, if the hydraulic machine is the rational model of the forms in which living matter is organized, and if we keep in mind that this concept suggests a strong analogy between capillary tubes and the canals that cross the body and establish communications with the outside through skin pores, it follows that it is proper to apply to canals the same explanations that have proved correct in capillary tubes. The theory of electricity provides an interpretation for the fact that the electrification of a fluid causes both evaporation phenomena and the acceleration of the outflow of liquids through thin pipes. The next step, then, is to find out whether these phenomena occur also in animals and vegetables, as could reasonably be expected. Such a test will afford a better knowledge of vegetables and animals and help in the planning of new experiments on both.

Nollet gives a detailed description of his first laboratory experiment, which goes as follows. Equal numbers of identical seeds are sown in two identical containers filled with soil of the same type. For a certain number of consecutive days container *A* is electrified at given times, while the other container is not subjected to the effects of the "electric virtue." The experiment begins on October 9, 1747, and the first control, performed on October 19, shows that in container *A* a great many shoots have appeared, each longer by a factor of four or five than the few shoots that have sprouted in the other container.

This phenomenon is systematically controlled by placing in the laboratory several pairs of containers and by using different kinds of seeds. The result of the test is unequivocal: "I have almost always observed a considerable difference between the electrified seeds and those that were not."[22]

Having ascertained the above beyond any doubt, Nollet turns to analogous experiments on animals. The analogy does not consist in trying to observe variations in the rate of growth but in using the "good inferences" derived from the fact that electricity accelerates the motions of fluids in thin pipes. "As a consequence of this truth," Nollet writes, "I have envisaged the pores that riddle the skin of animals as the ends of an infinite number of exceedingly small pipes, and perspiration matter as a fluid that tends to flow and whose egress might be helped or forced by the effluence of electric matter."[23]

The analogy used in the experiment on vegetable growth returns here as a prediction that can be subjected to strict experimental controls. If electricity increases the rate of perspiration, it is reasonable to expect that the weight of the electrified organism will "of necessity" decrease, which of course can be experimentally tested. This test had already been done by Boze with negative results. But if it is true that "in matters of physics the most respected authority is always subordinate to experience," Boze's test should be retested "to investigate such circumstances as might have escaped the first observers."[24]

The problem is to find a way to measure a constant and general effect, making sure that the experiment rests on sound bases and taking into account the possibility of experimental errors. The data will then be gathered into charts and discussed from various viewpoints. For instance, once the weight loss in all the organisms subjected to electric current has been ascertained, it might be possible to determine whether the loss is proportional to the surface or to the mass of the organisms.[25] This particular correlation, however, cannot be established from an analysis of the charts. What can be established is that the phenomena observed in animals and vegetables may be reproduced in the laboratory if, instead of electrifying the organisms directly, we place them near electrically charged bodies. This result confirms Nollet's theory that "there really exists *an inflowing matter* around electrified bodies" and that "electricity consists . . . *of two contrary and simultaneous motions of that matter we call electric.*"[26]

Figures 1, 2, and 3 show fairly well the train of thought Nollet follows in comparing the data and in drawing conclusions that can be interpreted as confirmations of the theory.

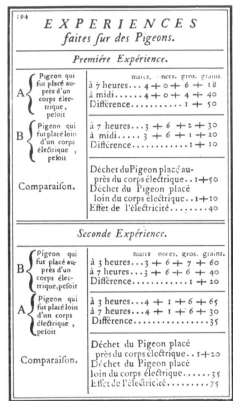
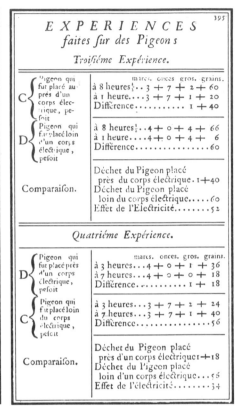

Figure 1   Figure 2

The data show some measurable effects of the electric virtue on animals (and on man). In addition to verifying the theory, the charts "unquestionably" prove that electricity could have practical applications. For instance, the cure of diseases due to an accumulation of harmful substances in the organism might be speeded up by applications of electricity. A case in point, according to Nollet, is the treatment of paralysis, as demonstrated by the experiments performed on patients at the Hôtel Royal des Invalides.[27] After protracted applications of electricity the patients showed on the average positive responses, although the statistics relative to the observed cases reveal some factors that are not completely clear. A number of observations and results, for example, appear to differ from country to country, as shown by the fact that doctors in France, England, and Germany have not been able to accomplish what has been done in Italy. Far from discrediting the experiments, these variations

Pigeons	L'animal étant électrifé.	L'animal étant placé près d'un corps électrique.
Expér. durée. produit.	produit.	
1.......5 heures...38 grains	...40 grains.	
2.......4.........55	57	
3.......4½.........50	52	
4.......4..........36	34	
Sommes des prod...179	183	
Termes moyens....44¾	.......45¾	

Bruants & Pinçons.	L'animal étant électrifé.	L'animal étant placé près d'un corps électrique.
Expér. durée. produit.	produit.	
1.......5 heures... 10grains	...11 grains.	
2.......4.........5	7	
3.......5.........6	9	
4.......5.........8	8	
Sommes des prod....29	34	
Termes moyens.....7¼	........8¾	

Figure 3

attest to the need for further research aimed at discovering the reasons for the statistical inconsistency. It is useless, Nollet writes, to take the attitude of those doctors who, being unable to cure paralytic patients with electricity, criticize the theory of electric phenomena. Nollet's own position is clearly stated in his answer to a surgeon of the Hôpital Général de Paris.[28] In a pamphlet published in 1747 Dr. Louis, "*chirurgien de la Salpêtrière*," claims to have applied the electric treatment suggested by Nollet to the paralytics of that hospital without any result whatsoever. The reason for this failure, Louis writes, is to be ascribed to the fact that "on the basis of all current knowledge of animal organisms, of the nature of the disease, and of electric power," Nollet's theory is patently wrong. Not true, Nollet retorts. The real problem is with Dr. Louis' ideas about the nature of the disease and with his opinions about electric phenomena and their theoretical interpretation. Since Dr. Louis has not understood the physical laws of electricity, his opinions are completely wrong. And with a distorted view of the the theory of electricity it is not proper to use the failure of the treatment to criticize the theory itself.

Theory and its fundamental assumptions are always the dominant theme in Nollet's work. In his opinion two conditions are equally necessary: theoretical deductions must be corroborated by facts carefully ascertained in the laboratory, and the formulation of the theory should not be undermined by any assumption that may be termed gratuitous—*precaria, & ex hypothesi petita.*

But how does the formulation of the theory of electricity actually come about? Even when the framing of the theory is done in accordance with norms that ensure its validity—observation of factual events, accuracy of measurements, anti-Cartesian tenets opposed to vortical motions, and so on—such faithful adherence to the norms is not sufficient to actually build the theory. Nollet does not restrict himself to a choice among available concepts and notions, nor does he accept that his choice should be regulated solely by methodological rules. He laboriously builds his theory (and defends it from criticism) by chains of inferences that draw upon different sectors of his culture—mechanics, electrology, theories of thermal and optical phenomena, along with opinions on the science of vegetable and animal organisms and evaluations of the import of mechanistic philosophy on the view of animals and plants as hydraulic machines. Nollet's reasoning proceeds within a vast dictionary, seeking arguments that will permit correlations to be found between distinct groups of phenomena and sets of observations and measurements performed on organic and inorganic matter.

The trouble with Nollet's theory is that within this vast dictionary and in the face of an ocean of data, Nollet is incapable of identifying a minimum set of problems that is solvable in its immediate context. Just the opposite: in Nollet's theory each specific solution must be immediately applicable to the whole ocean of data.

This is the salient aspect of Nollet's work and goes to prove that a scientist may be convinced of belonging to the ranks of the good Newtonians and yet be unaware of the danger involved in disregarding Galileo's warning on the need for "pruning down" nature, for isolating events and phenomena abstractly, for analyzing with theories all questions that are not too general.

Consequently, the task of the historian is not to look for mistakes in Nollet's theory, but to point out that weak interactions have a very great weight in that theory. This aim can only be achieved by taking into consideration what was accomplished in the field of electric research after Nollet. Subsequent developments are apt to shed light on what happened before. The correlations at work in Nollet's dictionary appear weak only after stronger correlations have been established within other dictionaries. Coulomb's physics, based on areas of mathematical rules and therefore bound by strong correlations, circumscribes Nollet's ocean of data and at the same time provides the historian with the elements necessary for reconstructing Nollet's theory of electric phenomena.

Electricity and Microphysics

One of the problems confronting Nollet is the relation between electric and thermal phenomena. Is there a connection between "electric matter" and the thin fluid supposed to be responsible for thermal phenomena? How does one relate fluids to the hypothesis that there is an identity of sorts between heat and motion? The most interesting aspect of this question is that Nollet does not formulate the problem in terms of a clear-cut choice between the hypothesis of heat as fluid and the hypothesis of heat as motion. In his *Leçons de physique expérimentale* he tries instead to establish some interdependence between the two theories.[29] This is important because he states in the *Leçons* that "the matter of electricity and that of fire" are the same fluid.[30] We have to be careful, however: fire is not heat. Fire is the cause "from which heat originates,"[31] and belongs to the "class of purely material bodies"—the "class of thin fluids."[32]

Now, when we say that heat and motion are synonyms, Nollet writes, we must also explain that this statement is not sufficient to identify the cause of the motion we observe in the various parts of a body. It is a true statement, but one that can be further analyzed. In sum, Nollet writes in the first pages of the first volume of the *Leçons*, what we have to identify in the theory is the prime physical cause.[33] Moreover, the prime physical cause must have the property of materiality: thus the theory must start not so much from a statement about the fact that the term heat denotes "the inner motion of the parts," but rather from a description of the nature of the thin material fluid responsible for that motion. "In the natural state, every motion, once started, slows down and finally stops being noticeable, communicating itself to a larger amount of matter. . . . Hence there must be an independent cause of the combustible parts . . . and this cause must be a substance."[34] Once the cause has been established, it must be related to the principles of mechanics to arrive at an explanation of the phenomena.

Two centuries later, we can certainly appreciate the complexity of the logical steps this explanation is based on. We can also appreciate the fact that Nollet is now in a position not to be forced to choose between the fluid model of heat and the model that depicts heat as a form of motion. By virtue of placing the prime cause in the class of thin fluids, his explanation shifts the problem from heat to fire and then resolves it in an attempt to unify thermal and electric phenomena. In short, Nollet proposes a fluid-based theory that postulates within itself the identity of heat and motion.

Nollet is also in favor of building a model of the fluid. The latter may be envisaged as a system of tiny hollow spheres bound by porous membranes and harboring in their interior "an ensemble of tiny particles of fire." Under normal conditions the particles cannot cross the membrane. But in the event of thermal perturbations of considerable intensity the particles of fire moving freely outside the tiny spheres assume a more violent motion and transmit this new state of excitation to the interior portions. This may eventually cause a break in the membranes; alternatively, the model may be refined by ascribing to the membranes thicknesses and pores of different sizes.

However, Nollet warns, the reader should not confuse a model with reality. The spheres are a product of our imagination, not of knowledge: they cannot be observed. "Needless to say, these tiny spheres full of fire . . . are by no means to be considered as something perceptible: even if they exist as portrayed by our imagination, these entities must be so minute that the smallest body observed through the microscope would contain a large number of them."[35]

These few remarks suffice to show how vast and complex was the general physical theory the Abbé Nollet attempted to construct, always seeking coherence and observational support. He extended his hypothesis of inflowing and outflowing matter to glasses, sulfurs, vegetables, cats, and man, and with the assumption of the "tiny particles of fire" attempted to investigate remarkably broad classes of things and events that could be tested in the laboratory. However, it would be a great mistake to regard Nollet's lengthy studies as a disorderly collection of deductions. In effect, from the Abbé's physical theory there emerges in full force the far-reaching idea that the cause of electric phenomena is a form of excitation of electric matter, which can be scientifically analyzed. The fluxes of inflowing and outflowing matter somehow give bodies the capacity to exhibit certain features, on the basis of which their degree of electrification can be investigated by theory. It is difficult to explain what it is that is introduced in electrified bodies: so difficult that it will take several decades to arrive by degrees at an explanation in terms of electric charge.

In the same years, around the middle of the century, a new hypothesis takes shape in the writings of B. Franklin, namely, the idea of something that instead of flowing in and out of bodies surrounds them in the form of atmospheres. Attractions and repulsions, which for Nollet represent only one of the relevant phenomena in the theory of electricity, in Franklin's hypotheses become a class of phenomena of the greatest importance. The distinction between classes of observable events acquires a history at every step in the development of the theory.

Franklin also starts from the same type of experiments as Nollet's: the power of points, the Leyden jar, the effects of electricity on man. But while in the 1746 *Essai* Nollet rejects the hypothesis of a "resinous" or "vitreous" electricity, Franklin, although in agreement with the idea that in electricity one deals with only one fluid, writes in 1747 that there may be two states of electrification.[36] The two states, positive and negative, indicate the electric condition of a body containing amounts of energy greater or smaller, respectively, than normal.

Naturally enough, Franklin observes the same type of phenomena investigated by Nollet: the behavior of electric fluxes of water, for instance. But where Nollet sees an acceleration of the flux, Franklin sees a complex of repulsions between drops.[37] Nollet speaks of a thin electric matter permeating bodies, and so does Franklin; but the two theories give different weight to the attractions and repulsions affecting this hypothetical thin matter. In Franklin's theory their importance is emphasized in connection with a model that will gain widespread acceptance in the second half of the eighteenth century, namely, the hypothesis that electrified bodies are surrounded by an electric atmosphere. Although a source of difficulties,[38] this hypothesis stands as a sort of bridge between macroscopic interactions and the interactions that the physics of those years is attempting to construct, at the microscopic level, between particles of thin matter and particles of ponderable matter. This bridge— or, if you prefer, this weak model—plays a decisive role in the investigation of what may reasonably be termed a very important phenomenon for the understanding of the laws of attraction and repulsion between charged bodies.

In the 1750s there is already a measurable distance separating Franklin's writings from the preface written by Francis Hauksbee for his book *Physico-Mechanical Experiments on Various Subjects,* published in 1709: "The General Laws of *Attraction* and *Repulse,* common to all Matter, have by the same Excellent Person been discover'd, and applied to Wonderful Purposes, in establishing the true System of Nature, and explaining the Great Motions in the World. But the *Nature* and *Laws* of *Electrical Attractions* have not yet been much consider'd by Any."[39]

In other words, if we start from Franklin, the deductions of Hauksbee or Nollet become much clearer: Franklin's dictionary helps us to analyze prior dictionaries and to reconstruct their inner logic. In addition, this analysis clarifies the meaning that the various scientists give to the term "experience." What they call experience and attempt to condense in tables of numbers, sets of measurements, and methodological definitions, is not something that accumulates over the decades waiting for the explicative act of a genius. What we see over the decades is, rather, a modification of the so-called empirical basis, and such modification is brought about by theories that continually act on data through reclassi-fication and reinterpretation, addition, elimination, and prediction. We do not mean by this that the observed effects of electricity on paralytics stop being observable. What we do mean is that such effects are no longer considered basic to the formulation of the theory of electric phenomena. In the late eighteenth century the growing interest in animal electricity is a result of just such modifications.

Thus, in the transformation of the empirical basis physicists do not confront an experience that is some sort of mental projection of the theories. The ever-changing ocean of data is not the offspring of an array of conjectures or of a clash between theories. The instability of the empirical basis is a result of the variations in the answers given by nature to scientists who probe her with different theories.

When Nollet and Franklin observe the power of points and draw different conclusions from their observations, we are not watching a drama in which intense observation of the points alters their power. Such creative interaction between observations and the world is only a dream of magic, an epistemological rephrasing of the belief that parading a wooden statue will cause rain to fall on the fields.

The thesis that experience is unstable challenges the credibility of any historical reconstruction of scientific inquiry based on the notion that science is an accumulation of practical cognitions. The instability of experience, a verifiable result of the Galilean proposition that theories violate the senses, by no means supports the thesis that nature tamely fashions herself according to our ideas and actions. In actuality, nature resists and rebels and in so doing behaves very improperly by the standards of idealistic doctrines. It is not by chance that there are profound similarities between the history told by a strict "internist" and the philosophy taught by an idealist: the former confines real history to the footnotes; the latter does the same with nature.

Cavendish and the Reconstruction of the Empirical Fabric

In the foregoing paragraph we mentioned differences between the various dictionaries. Let us take a closer look at this problem.

In 1772 Henry Cavendish publishes a work entitled "An Attempt to Explain some of the Principal Phenomena of Electricity, by Means of an Elastic Fluid."[40] As usual, the question of method comes up immediately. In the very first lines we read, "The method I propose to follow is, first, to lay down the hypothesis; next, to examine by strict mathematical reasoning, or at least, as strict reasoning as the nature of the subject will admit of, what consequences will follow from thence; and lastly, to examine how far these consequences agree with such experiments as have yet been made on this subject."[41]

Cavendish's hypothesis sets a particular form of the interaction between matter and electric fluid at the center of the problem of electricity. "There is a substance, which I call the electric fluid, the particles of which repel each other and attract the particles of all other matter, with a force inversely as some less power of the distance than the cube: the particles of all other matter also, repel each other, and attract those of the electric fluid, with a force varying according to the same power of the distances."[42]

It will be noted that it is not a simple hypothesis. The statement that this particular form of interaction is of the type $1/r^n$ and that some conditions are to be set on n does not make clear the fundamental difference between a particle of electric fluid and a particle of matter.

Cavendish is aware of the problem inherent in his hypothesis and attempts to overcome this difficulty by confronting it head on. The electric fluid permeates material bodies and in so doing interacts with their particles not only in proportion to some power of the distances but in proportion to the "weight" of each interacting particle. From a macroscopic point of view, this may be explained by the fact that the weight of the fluid in a body must be far less than the weight of the body itself: the mass of a particle of electric fluid must therefore be exceedingly small.

However, "the force with which the electric fluid attracts any particle of matter must be equal to the force with which matter repels the particle." If this condition of equilibrium is violated the body becomes "electric."

Thus the hypothesis must be further qualified to state that the intensity of the forces pertaining to the fluid is much greater than that of the forces inherent in ordinary matter.[43]

This done, the hypothesis permits us to classify bodies into three groups: "All bodies in their natural state, with regard to electricity, contain such a quantity of electric fluid interspersed between their particles that the attraction of the electric fluid in any small part of the body on a given particle of matter shall be equal to the repulsion of the matter in the same small part on the same particle. A body in this state I call saturated with electric fluid: if the body contains more than this quantity of electric fluid, I call it overcharged: if less, I call it undercharged. This is the hypothesis."[44]

How can consequences be drawn from this hypothesis? The 1772 work leaves no doubt on the subject. The consequences of a hypothesis are obtained by formulating lemmas and discussing corollaries in which reasoning proceeds on a mathematical basis.

After enunciating the hypothesis, Cavendish states lemma I: "Let EAe represent a cone continued infinitely; let A be the vertex, and Bb and Dd planes parallel to the base; and let the cone be filled with uniform matter, whose particles repel each other with a force inversely as the n power of the distance. If n is greater than 3, the force with which a particle at A is repelled by $EBbe$ or all that part of the cone beyond Bb is as $1/AB^{\,n-3}$."

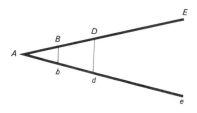

Starting from lemma I and two other lemmas relative to those cases in which $n = 3$ and $n < 3$, respectively, Cavendish enunciates a corollary:

> It is easy to see from these three lemmata, that, if the electric attraction and repulsion had been supposed to be inversely, as some higher power of the distance than the cube, a particle could not have been sensibly affected by the repulsion of any fluid, except what was placed close to it. If the repulsion was inversely, as the cube of the distance, a particle could not be sensibly affected by the repulsion of any finite quantity of fluid, except what was close to it. But as the repulsion is supposed to be inversely as some power of the distance less than the cube, a particle may be sensibly affected by the repulsion of a finite quantity of fluid, placed at any finite distance from it.[45]

From the three lemmas and their corollary Cavendish logically proceeds to enunciate a definition and two propositions, and then continues to build the theory, formulating problems and their solutions and positing additional lemmas, definitions, and corollaries.

There is fundamental difference between Cavendish's approach and Nollet's. Cavendish adopts well-defined rules and acknowledges his debt to Aepinus's strongly mathematical approach. His theory defines the questions to be asked of nature, thus reducing the number of phenomena that have to be taken into account. Nollet's empirical ocean disappears and theory erects conjectures and lemmas on well-circumscribed experimental grounds.

Nollet's problem of attractions and repulsions becomes with Cavendish a much better-defined problem. And this more precise definition contains within itself the possibility of a solution, since Cavendish establishes that the law of electric force is in the form $1/r^2$ to a very good approximation.[46] Whereas for Nollet the question of attraction is connected with methodological principles of a Newtonian stamp, with Cavendish it becomes a mathematical problem. As such, it is no longer discussed simply in terms of a clash between the Cartesian doctrine of vortices and the Newtonian doctrine of action at a distance, but in the much more limited and precise context of algorithms and demonstrations.

Of course it might still be argued that once the individual differences between the theories of Nollet, Franklin, and Cavendish have been duly noted, a genuine philosophical issue still remains to be considered, namely, the mechanistic explanation. To this we would answer that after much reflection on the various aspects of this issue nobody has been able to derive from it the body of correlations Nollet was trying to establish between the laws of hydrodynamics, the electrification of cats, and the treatment of paralysis. Nor is there anybody who after pondering Cavendish's method has been able to enunciate even one of Cavendish's mathematical problems. Such failures are bound to mean something— which is not necessarily a devaluation of method. Rather, they should make us reflect on the interactions of methods and theories that take place within historically determined dictionaries. In addition, they should inspire greater caution in those who, for one reason or another, believe individual theories to be mere formal appendages to the method. One of the most insidious ways to dissolve objective knowledge into philosophy is precisely to turn scientific enterprise into a purely methodological enterprise.

Return to the Boltzmann–Kelvin Problem

It is our hope that the result of our wanderings through the history of physics will be more than just the loss of that naiveté that characterizes the view of classical physics as a science dominated by mechanistic philosophy. The path we have followed thus far has perhaps given us the tools to begin anew the exploration of the origins of the second scientific revolution, having marked on the map the obstacles and prejudices we have inherited from the great controversy over the natural sciences that troubled the first forty years of our century.

Let us return now to the end of the nineteenth century to find, once again, Kelvin's dictionary and Boltzmann's battle. This will also give us the opportunity to ponder the fate of those strange things that physicists called caloric and ether, and for which they gave or sought experimental proof. The elimination of these two entities of universal dimensions is a very interesting process, in that it shows that the idea of matter with which physicists are concerned must always be distinguished from the idea of matter entertained by philosophers. On the possibility—or necessity—of this distinction revolves not only Ostwald's philosophical denial of matter in general, but the centuries-old fight for a rational view of knowledge.

It is not an accident—nor a fact wholly analyzable in terms of psychology or sociology of science—that in our century the new sciences have been nourished by the great debates between Bohr and Einstein, Dirac and de Broglie. The so-called crisis in physics, which according to some philosophers was brought about by the collapse of classical mechanism and of a vulgar metaphysical materialism, is nothing but a philosophical metaphor meant to cover up philosophy's refusal to acknowledge the changes effected by the second scientific revolution in the realm of theory. Philosophy becomes obsolete when it refuses to adapt itself to new scientific discoveries.

The New Physics

What is meant by the statement that most physicists are apt to hold a realistic view of the relationship between their science and the world? One answer might be the following: most physicists hold a realistic view in the sense indicated by Galileo's splendid thought, "Our discourses must relate to the sensible world and not to one on paper."[1]

However, we must consider a particular characteristic of the physical theories developed after the Galilean revolution. Because of the increasingly important role played by mathematics, theories gradually evolved in directions that we can no longer explain by remaining entrenched in philosophical realism. The physicomathematical theories of the second scientific revolution did not simply achieve a clarification of specific segments of the world. They developed in such a way as to contribute to our understanding of all previous theories.

In sum, the new sciences of the second scientific revolution hold discourses about the sensible world and about themselves.

We do not mean by this that the second scientific revolution runs counter to Galileo's aspiration to know the world rationally by means of sensible experiments and theories that violate the senses. The reason for mentioning Galileo's distinction between the paper world and the real one is to point out that such a distinction is no longer adequate for a rational view of the relationship between science and nature.

With the new awareness of the futility of assuming the world to be endowed with the same structure in the infinitely small and in the infinitely large, the new physical sciences found themselves confronted not only with a world full of surprises but with objective problems originating in the theories themselves.

Two examples will help us to make our point. The first concerns so-called classical mechanics. The generalizations of mechanics elaborated by Hamilton and Jacobi in the first half of the nineteenth century were not merely extensions of Lagrange's mechanics. Discussing the motion of bodies according to Hamilton's scheme is not the same as discussing the motion of bodies according to Lagrange's scheme. And the differences are not just formal since starting from these differences one arrives at different descriptions of the world. This fact was already apparent to physicists like Maxwell, who adopted Lagrange's mechanics knowing full well that it would free them from the need to make statements about unobservable properties. While outlining a dynamical treatment of the electromagnetic field, Maxwell wrote that since "the nature of the connections of the parts . . . is unknown to us" and since there is always an infinite number of solutions to the problem of making a mechanical model of such connections, Lagrange's mechanics is indispensable in that it provides "dynamical methods of investigation which do not require a knowledge of the mechanism of the system."[2] When a scientist makes a specific theoretical choice he describes the world in a certain manner (and in no other); at the same time he realizes that he is speaking about a problem whose object is the relationship between different theories.

The second case concerns the relationship between classical mechanics and quantum mechanics. The very special nature of this relationship, and the body of problems it raises, involve fundamental questions that L. Landau and E. Lifshitz have summarized thus: "Quantum mechanics occupies a most original position in the ranks of physical theories—it contains classical mechanics as an extreme case and, at the same time, it needs this extreme case in order to be founded."[3] In the passage from classical to quantum mechanics problems arose that concerned not only the strange objects of the microstructure of matter but the very structure of the theories.

In short, the problems of the new physical sciences concern both the sensible world of Galileo and the new relationship between theories. The latter question cannot be resolved by assuming that it is a particular concern of the paper world: the analysis of the structure of empirical theories is concerned with both their form and their content.

It will thus be useful to examine topically a few problems relevant to our view that a mathematical physical theory is built in such a way as to discuss sensibly both material objects and theoretical questions.

Ether and Caloric

Let us start from the assumption that the purpose of a mathematical physical theory is to determine a number of parameters of certain objects.

The objects in question are of various kinds. In the mathematization of astronomy, for example, theory discusses things, such as planets, that exist independently of us. Can the kind of reality that in this example we ascribe to the term "thing" be safely ascribed to other things discussed in other theories? The question is proper and the answer, as we all know, far from simple. In the nineteenth century it took decades of laborious study to formulate powerful, mathematically-based theories capable of describing the properties of the two entities that were considered the largest, most extended "things" in the whole universe: caloric and ether. In both cases further development of the theories—thermodynamics and electromagnetism—led to a realization that those two entities were not things in the accepted sense of the word, that is, as the word was used in current parlance. An analogous situation had already occurred at the end of the eighteenth century with the disappearance of another "thing" of universal dimensions: the chemists' phlogiston.

The theoretical demolition of such objects was naturally bound to cause controversy and heated debates among physicists, chemists, and philosophers. The disappearance of things like ether and caloric did not seem too disturbing to the scientists whose judgment was determined by the realism affecting their philosophical conception, or spontaneous philosophy, but for many others it represented a mortal blow to the materialistic approach. The former often found a modicum of philosophical comfort in the realistic point of view that the loss of caloric from the field of science had been brought about by crucial experiments. The decision to remove caloric from the universe did not appear to them as a problem affecting theory's ability to describe things, but rather as a triumph of experience and therefore as a triumph of man's ability to reason sensibly about what exists. Of great comfort to the physicists of the second half of the nineteenth century was a paper that asserted that the passage of physics from caloric to thermodynamics was a victory of experimental reasoning.[4] The same argument was used for the ether: its disappearance and the genesis of special relativity would have been brought about by the Michelson–Morley crucial experiment—a statement we still find today in some authoritative textbooks of physics.

However, the optimism evinced by this kind of explanation is quickly dampened by historical research, which shows that the elimination of the object ether and the inception of special relativity were not the result of a crucial experiment but of Einstein's theoretical studies on certain asymmetries in Lorentz's theory.[5] Historical research also proves that the elimination of the object caloric and the beginning of thermodynamics were not due to the presumed crucial experiments by H. Davy and B. Thompson but to a radical reframing of the theories formulated by Laplace and Poisson and to the realization that such reframing was necessary to overcome theoretical contradictions. In both cases the relation between theory and experiment was far more complex than it appears to those who believe that a series of crucial experiments constitute the prime mover in physics (and in the history of physics). And in both cases the objects in question—ether and caloric—were eliminated because the development of the theories enabled scientists first to identify in them the "theoretical objects" and then to demonstrate that these theoretical objects were a source of contradictions and paradoxical consequences.

The attitude of the scientists who played leading roles in these events appear to have been shaped by two different drives. On the one hand, as physicists, they formulated new concepts aimed at solving a set of problems concerning the world—Does caloric exist?—and formulated them in such a way as to bring to light the contradictions undermining the old theories. On the other hand, as proponents of philosophical views, these same scientists tried to justify the formulation of new concepts and to frame them in such a way as to safeguard certain methodological norms. This brings to mind Einstein's suggestion that we ought to look at what physicists do rather than at what physicists say they do. But even if we accept Einstein's suggestion, the problem still remains that nineteenth-century mathematical physical theories contain statements bearing on things.

To summarize: the historical evolution of mathematical physical theories leads to the construction and demolition of significant objects and, concurrently, to the identification of problems that have to do with the very structure of the theories; at the same time, the philosophy of each scientist forces him to find explanations by the eclectic use of philosophical concepts—experience, crucial experiment, matter. To Einstein's suggestion should thus be added the polemic statement made by Faraday in 1857 against Newtonian dogmas, "I am unable to define what is metaphysical in physical science."[6]

The philosophy of scientists like Faraday is an oscillating system rather than a static hierarchy of concepts. In the process of restructuring a theory and of disproving what was previously believed to be empirical evidence we often see strategies that support Faraday's point: no clear demarcation line is ever drawn between physics and metaphysics; moreover, we find statements that "even an obscure and distorted vision is better than none."[7] The philosophical illusion that prompts some historians to seek in all theories a nucleus of laws to be defended at all costs fails to find support in the scientists' own philosophies. These historians cannot accept the fact that physicists like Faraday, Laplace, Fourier, and Maxwell were not slaves to a philosophical system given once and for all but held philosophical views that moved back and forth within their own evolving dictionaries.

Laplace's Caloric

Let us return now to the object caloric. Before 1830, the term caloric is generally understood—with some notable exceptions—to mean a thing that is present in the whole universe, permeates all bodies, and moves between one body and another at a speed of the same order of magnitude as the speed of light. All thermal phenomena are a consequence of the behavior of this thing and a good many French mathematical physicists are engaged in the demanding work of rationalizing its behavior.

The research approach followed by Laplace and Poisson produces a masterpiece of mathematical rationalization of the caloric theory. In particular, the author of the *Traité de mécanique céleste* develops the theory to the point of achieving an experimental demonstration of the existence of latent caloric, obtaining a surprising degree of empirical verification for certain theoretical predictions.[8]

Leaving aside the system of deductions and physical interpretations in Laplace's theory, the experimental confirmation of the existence of latent caloric may be summarized as follows.

Caloric is a discrete fluid (consisting of distinct elements) that interacts with the molecules of a body. Because of this interaction, caloric exists in two different physical states: as latent caloric surrounding the single molecules, of which direct experimental confirmation cannot be obtained; and, in the spaces between molecules, as thermal radiation consisting of elements of free caloric. In the latter state the fluid can be observed, because theory defines a measure of temperature as a measure of the density of thermal radiation.

Up to now the term "latent caloric" denotes a physical state that is not observable by direct temperature measurements. According to Laplace, however, it is possible to plan an indirect observation of latent caloric.

This indirect observation is connected with a classical problem of theoretical physics, namely, the problem of finding a formula for the speed of sound in gases that will improve the Newtonian formula, whose predictions differ substantially from actual measurements. Laplace has at his disposal a theory of caloric that describes in a general way the action caloric can exert on a given point inside a cylinder full of gas. According to Laplace, the problem can be reduced to that of setting certain specific conditions. These conditions are expressed by a mathematical formula that relates the densities of the gas (ρ and ρ') and the quantities of caloric (c and c') at two different points on the x-axis of the cylinder separated by a distance s:

$$\rho c' = \rho c + s \frac{\delta \rho c}{\delta x}.$$

The relation is further defined by introducing a quantity θ, the meaning of which remains to be specified:

$$\frac{\delta \rho c / \delta x}{\rho c} = (1 - \theta) \frac{\delta \rho / \delta x}{\rho}. \tag{A}$$

Laplace's first step relies on the validity of the fundamental principle of the caloric theory, which states that the sum of free caloric c and latent caloric i is a constant. It follows that if latent caloric is always zero, free caloric is a constant. Hence, from (A) one derives $\theta = 0$.

On the other hand, from (A) and from the equation describing wave propagation in the gas, the velocity of the wave is found to be a function of θ:

$$V = \sqrt{2V_{\mathrm{N}}(1 - \theta)}, \tag{B}$$

where V_{N} is the velocity according to Newton. The new equation deduced by Laplace agrees with the experimental data only by assuming that θ is not zero, which implies that i is not zero.

"This experiment thus proves," Laplace writes, "that there is a latent heat i in gas molecules."

It might be remarked at this point that Laplace's whole approach is more than a little ambiguous. Although close to nothing is said about θ, except for what is said in (A), the factor θ is in effect the cardinal point of the entire procedure. Laplace tackles this problem by bringing into play other portions of the theory.

The absolute heat $c + i$ is a function of the density ρ of the gas, of the pressure P, and of the temperature μ:

$c + i = \psi(P, \rho, \mu)$.

Now, taking into account (A), the principle of constancy of $(c + i)$ and the laws of gases—which express the pressure as a function of ρ and c and are experimentally confirmed by Mariotte, Dalton, and Gay-Lussac, Laplace derives the following relation:

$$1 - \theta = \frac{1}{2} \frac{c_p}{c_v} ,$$

where c_p and c_v denote specific heats for a constant pressure and for a constant volume, respectively.

Hence (B) becomes

$$V = V_N \left(\frac{c_p}{c_v} \right)^{1/2} . \tag{C}$$

Relation (C) may easily be subjected to experimental tests, which it successfully withstands, as shown by the charts published by Laplace. Actually it is so successful that it can also reconcile mathematical physical predictions and the measurements of the specific heat of air as a function of pressure published in 1813 by Delaroche and Bérard and considered unimpeachable at the time.[9]

Since the theory appears to be coherent and enjoys the support of laboratory tests, it becomes well-nigh impossible to raise doubts about the existence of caloric.

Laplace's demonstration is part of a theoretical construction that originates from a set of assumptions concerning the interactions between the discrete structure of matter—the ponderable molecules, and the discrete structure of the fluid—the elements of caloric. In this theoretical construction a leading role is played by several inferences that draw their validity from the calculus of probability coupled with the mathematics of rational mechanics. Thus it would be incorrect to regard the theory of gases and caloric as a mechanical theory built according to the norms of a mechanistic and deterministic methodology.

To discuss Laplace's approach in terms of deterministic mechanism simply means to discuss something that does not exist in his approach. It should also be noted that an analysis of Laplace's approach entails not only a study of its form but a study of its content: the demonstration of the existence of caloric is no mere problem of linguistics or methodology.

Fourier's Caloric

In the same years J. B. Fourier develops a different approach from that of Laplace. In the *Théorie analytique de la chaleur* Fourier formulates a generalization of the theory of the series known by his name and frames a theory of thermal conduction, but his mathematicophysical tools are not concerned with analyzing the structure of caloric. According to Fourier, all the phenomena of thermal conduction are to be brought into the class of the solutions of a differential equation, and the form of this equation does not depend on a set of assumptions concerning the interaction of matter and caloric, but on a general principle and on deductions regulated solely by the type of calculus one chooses.[10]

When Fourier asserts that the theory of heat is independent of mechanics, he means actual independence. For Fourier the validity of his theory does not depend on the possibility of elaborating an interpretation of it that will conform to Laplace's fluid theory or to the wave theory, nor on the highly questionable possibility of deducing its equations from the principles of dynamics.

Fourier derives the form of the differential equation from the demonstration of a theorem:[11] "Let us assume that the different points of a homogeneous solid of any given shape have initial temperatures that vary in time because of the mutual action of the molecules, and that the equation

$$v = f(x, y, z, t)$$

represents the states of the solid; it can be demonstrated that the function v of four variables necessarily satisfies the equation

$$\frac{\delta v}{\delta t} = \frac{k}{CD} \left(\frac{\delta^2 v}{\delta x^2} + \frac{\delta^2 v}{\delta y^2} + \frac{\delta^2 v}{\delta z^2} \right).$$

The demonstration assumes that what is meant by mutual action of the molecules has been explained. Whereas in Laplace and Poisson the meaning of mutual action is illustrated by a model of the emission and absorption of thermal radiation by each molecule, in Fourier it is based on the general principle that the amount of heat a given point of the homogeneous, isotropic solid receives from another point is a function of the distance and of the difference in temperature. Fourier eliminates models for the simple reason that he can discuss the problem of flux mathematically. And all this is neither simply a matter of form nor a banal question of adherence or nonadherence to a mechanistic view of the world.

Trying to unify Laplace's and Fourier's mathematical physics by means of considerations involving mechanism results in nothing more than a literary discussion of a meaningless question that can be summarized as follows: What is the deterministic content of a differential equation? The number of philosophical theses that have been written to clarify the meaning of such a silly question is truly astounding. What is not at all amazing, instead, is that Ampère should have gone straight to the heart of the matter in the 1835 essay we mentioned earlier.[12] Using historically given notions, Ampère understood the paradoxical situation that arises in physics when both Fourier's theory and the wave theory of heat and light are assumed to be valid. Again, this is not a mere formal question, in the sense that it has a direct bearing on what physics has to say about the world.

Thus the task of the historian is not only to stress the importance of a paper by Ampère that has not always received the attention it deserves but, more important, to point out that objective problems get treated by historically given theoretical notions.

However, if the historian is convinced that the differences between the physics of Ampère, Fourier, and Laplace are essentially minor differences in view of their presumed common adherence to one method, one mechanistic philosophy, and one body of dogmas, then his work becomes meaningless and is reduced to inconclusive prattle.

The question of the relationship between the theories we have discussed and the specific ways in which these theories investigate nature and themselves—their own objects and contradictions—is particularly interesting because it shows that seventeenth-century mechanism is in fact put to rest in the first years of the nineteenth century. The burial services are not attended by the heirs of a mechanistic methodology absurdly unaware of the new objective problems, but by a group of scholars who, by working rationally on the new problems and by formulating new solutions, initiate the second scientific revolution.

The Objects of the Physical World

Mathematics' massive intervention in the theory of thermal phenomena raises new problems that concern both the structure of the theory and those portions of it that hinge on the existence of an object of universal dimensions. The problem of caloric becomes more and more a problem of compatibility between mathematical statements and propositions that interpret phenomena that are verifiable in the laboratory. It is a many-faceted problem since it can be maintained that the art of experimentation is guided by theory.

The increasing complexity of the relationship between a growing theory and constantly improving measurements explains, among other things, the relatively long period of time it takes scientists to abandon definitively the concept of caloric. Nothing happens instantaneously in this historical process. But the lack of instantaneous developments does not imply continuity in the process itself. The long transition phase that leads up to the founding of thermodynamics proceeds neither by skips and jumps nor according to linear patterns. Although it may not follow the traditional schemes based on continuity or discontinuity, this transition phase cannot be reduced to an intermezzo dominated by strategies without rules. The initial phase is not homogeneous, in the sense that opposing mathematicophysical theories coexist in it. The controversy between Fourier, Laplace, and Poisson does not suddenly break out with the publication of the *Théorie analytique de la chaleur* but is already evident in the second decade of the century. Moreover, the controversy is not confined to the question whether thermal phenomena can be reduced to the principles of dynamics. From the very beginning it concerns mainly the criteria for choosing the mathematical approach best suited to the formulation and solution of the problem of thermal conductivity: To what extent does the theory of Fourier series guarantee the validity of the physics of thermal phenomena? The initial phase, in other words, is already rich with inferences that bear on a vast array of problems. In the transition period, which reaches peaks of particular intensity between 1830 and 1847, theoretical and experimental problems stemming from basic contradictions in theoretical physics are examined and identified, analyzed and enunciated. The conflict between the wave theory and the caloric theory is clear-cut, but the battle is waged with systems of rules that are securely anchored in mathematics.[13] The controversy, in sum, is based on arguments that, however much they diverge, do not systematically violate the

rules on which they rest. The clash between the theses of Lamé,
M. Melloni, Poisson, Thomas Thomson, and Ampère implies neither a
fanatical adherence to a scientific paradigm nor useless concessions to
mystical attitudes. The question whether caloric does or does not exist is
not decided by personal tastes but by the possibility of formulating the
problem correctly and of solving it according to the rules of mathematical
physics. The psychology of research may explain the motives that prompt
a scientist to adopt a newly developed theory but has no bearing on the
actual process of development of the theory itself.

In support of this statement we will examine the controversy that arose
around Joule's experimental results and their anticaloric implications. It
is often said that Joule's data were long opposed by a large and powerful
group of English scientists, whose stubbornly negative attitude was
mainly due to their methodological conservatism, staunch opposition to
any innovation, and the desire to hold on to their own ideologies. Let us
then consider Joule's papers. Of particular interest to us is the essay "On
the Mechanical Equivalent of Heat," published in 1850 in *Philosophical
Transactions*.[14] The tables of data presented by Joule concern measurements of changes in temperature each lasting over half an hour and
entailing corrections of the order of a millionth of a degree on readings of
the order of $0.575250°F$! It is not necessary to invoke ideologies to find a
few reasons for disagreement. All of this probably explains why as late as
1848 Kelvin still owns to some doubts: "The conversion of heat (or *caloric*)
into mechanical effect is probably impossible; certainly undiscovered."
And this is true, he adds in a footnote, although "a contrary opinion . . .
has been advocated by Mr. Joule of Manchester."[15]

The key to understanding the perplexity shared by most physicists lies in
Kelvin's reflection at the close of the abovementioned paper: "It must be
confessed that as yet much is invested in mystery with reference to these
fundamental questions of natural philosophy." What is involved here is
not a millionth of a degree Fahrenheit or the ideology of conservative
physicists: it is a complex of questions affecting the very foundations of a
theory that has been argued, defended, and fought over for more than
seventy years and that is still able to correlate vast classes of phenomena.

On the other hand, the validity of Joule's theses cannot be accepted solely
on the basis of long sets of measurements that supposedly speak for
themselves. In fact, there must be conducted a radical criticism of one of
the pillars of the caloric theory—a criticism so radical as to devastate the

theory, which after all has given precious service and has enabled Kelvin, after the introduction of Sadi Carnot's propositions, to enunciate to a first approximation the concept of absolute thermometric scale. What would be left of this grand edifice if the object caloric were to be replaced by the theoretical idea that heat is molecular motion?

To the problem identified by Ampère in 1832 and 1835 is now added an even more serious set of problems. They come to light with the initial successes of the wave theory of heat and stem from the contradiction the new theory runs into as it attempts to incorporate both Joule's ideas and Carnot's—ideas that appear incompatible.

In 1850 W. J. M. Rankine proposes what appears to be a satisfactory solution based on a model of the structure of matter.[16] According to Rankine's hypothesis each atom consists of a nucleus surrounded by revolving particles. Thermal phenomena would then be explained by the fact that the quantity of heat is the live force of the motion of the particles around the nuclei. An additional assumption is required to account for light and heat radiation: the medium that transmits light and radiating heat consists of the nuclei of atoms vibrating independently, or almost wholly independently, of their atmospheres. According to Rankine, the microphenomena at the nuclear and particle level may be described by a variant of the hypothesis of molecular radiation on which caloric theories were based. *Absorption* is produced by transferring motion from the nuclei to the atmospheres (of the particles), *irradiation* by the transfer of motion from the atmospheres to the nuclei.

In this model the emphasis is on the innovative principle that heat is a form of motion rather than an object, which will now make it possible, in Rankine's opinion, to deduce the two laws of the new mechanical theory from the principles of mechanics.

The hypothesis of molecular vortices promises great things. Moreover, as Rankine remarks in 1854, it can be defended from the standpoint of orthodox Newtonianism: "Of this hypothesis, as of all the others, neither the veracity nor falseness can be demonstrated: it is simply probable in proportion to the number of facts with which its consequences are in agreement."[17]

The great promises are not kept, however. The contradiction between the two principles of the new theory of heat does not simply go away while men wait for a formally correct deduction of the two principles of mechanics. And the problem is no closer to a solution if one accepts the view, first expressed by Joule in 1845, that Carnot's axiom should be abandoned together with the belief in the material existence of caloric. In 1849 Joule's position is criticized by Kelvin on general grounds.[18] After the discussions held at the Oxford meeting of 1847 Kelvin has begun to revise his own ideas on the theory of heat. He now accepts Joule's principle but does not agree that the difficulty will disappear by abandoning Carnot's axiom. What is necessary, in his opinion, is "an entire reconstruction of the theory of heat from its foundation." Kelvin sums up the situation with these words:

When "thermal agency" is thus spent in conducting heat through a solid, what becomes of the mechanical effect which it might produce? Nothing can be lost in the operations of nature—no energy can be destroyed. What effect then is produced in place of the mechanical effect which is lost? A perfect theory of heat imperatively demands an answer to this question; yet no answer can be given in the present state of science. A few years ago, a similar confession must have been made with reference to the mechanical effect lost in a fluid set in motion in the interior of a rigid closed vessel, and allowed to come to rest by its own internal friction; but in this case, the foundation of a solution of the difficulty has been actually found, in Mr. Joule's discovery. . . . Encouraged by this example, we may hope that the very perplexing question in the theory of heat, by which we are at present arrested, will, before long, be cleared up. It might appear, that the difficulty would be entirely avoided, by abandoning Carnot's fundamental axiom; a view which is strongly urged by Mr. Joule. . . . If we do so, however, we meet with innumerable other difficulties— insuperable without further experimental investigation, and an entire reconstruction of the theory of heat from its foundation. It is in reality to experiment that we must look—either for a verification of Carnot's axiom, and an explanation of the difficulty we have been considering; or for an entirely new basis of the Theory of Heat.[19]

This passage has a precise meaning. When a theory is in the formative stage, in order to choose the best answer to a question concerning an object we have to reflect on the reasons why that question represents a serious difficulty. Consequently, the answer no longer concerns the question alone but bears on the very foundations of the theory. No methodological decree can ever guarantee the absolute validity of the foundations and therefore there are no shortcuts to the formulation of a theory, only long tortuous routes. It is true that the dictionary may suggest either one route or another and help in weighing an uncertain idea against a verifiable argument; but once started on a given route, the theory proceeds by steps whose validity is not an immediate function of the whole dictionary but only of a subset of its rules. And therein lies the relative autonomy of scientific inquiry.

Thus the propositions of a mathematical physical theory with which we describe objects or things are not always and uniquely the product of the action whereby one or more scientists leave the realm of physics and enter surreptitiously into that of philosophy. Apart from the very controversial and much debated question whether philosophy has a domain of any kind (and therefore any barriers separating it from physics), the fact remains that physical theories build objects that belong to the physical world. Both theoretical concepts and sensible experiments are used to build and demolish such objects. And since theory indeed violates the senses, the existence of the objects that—for a time—belong to the physical world is not a question that can be resolved by pointing one's finger at them. The object Mars and the object caloric are not different from each other because the former can be pointed at in the starry sky while the latter cannot. As far as the physical object Mars and the physical object caloric are concerned, scientists can only point at a library of textbooks and papers on celestial mechanics and at a library of textbooks and papers on the theories of thermal phenomena.

A library, not a manual. Even an object as massive and conspicuous as Mars has been subjected to an amazing amount of theoretical violence by virtue of being a physical object, that is, an object belonging to the world of physics. Hence the modifications of "Mars" can be pointed out only by showing the development of the theories regarding Mars, just as the modifications of "caloric" can be pointed out only by showing the development of the theories of thermal phenomena.

The inability to distinguish between the objects of the world of physics and the stones used in building houses is the true source of the philosophical dilemma that is periodically solved by proclaiming the disappearance of matter.

Galileo's epigram about the real world and the world on paper has to be gradually reinterpreted through the new cognitive elements that the second scientific revolution has brought to light and continues to deepen. There is no more room, save in literary circles, for disputes over the absolute concept of matter and for debates over the percentage of mechanism in a certain paper by Maxwell or on the deterministic content of a certain set of differential equations.[20] If room is left for such meaningless topics, the philosophical–scientific discourse will be turned into a metaphor of real problems, which is precisely what happens, for instance, when a barrier is erected between the "matter" of the philosopher and the "physical world" of the scientist. This barrier—which many consider desirable as an insuperable limit to the cognitive activity of science—in effect creates a mythological no man's land where the so-called intuitions of the deepest truths, the blinding revelations, the psychologic dramas of the Spirit, and occasionally the biographies of great scientists are played out.

The world of physics and the objects pertaining to it may be defined as the body of statements—organized into theories—with which we describe the world we are part of and conditioned by. Leaving to others the task of raising doubts about the existence of the objective world, the physical sciences ask sensible questions about Mars or caloric and interpret the answers by appealing to areas of rules. The structure of the questions and answers is historically determined, since the areas of rules are historically determined and evolve in a real historical process that involves all the dictionaries. Vesalius's or Millikan's experiments are not historical examples of the absolute concept of experiment; nor are Galileo's explanation of the fall of bodies or the quantistic explanation of atomic structure historical examples of the absolute concept of scientific explanation. Real historical processes, in sum, are not repositories of edifying examples. A theory of knowledge that continues to cling to absolute ideas is nothing more than a catechism in search of a false history.

Ostwald's Mechanism and Kelvin's Clouds

The new sciences of the second scientific revolution taught philosophical realism some harsh lessons. Once again, however, these lessons were interpreted with different dictionaries and led to different scientific-philosophical reactions. Ostwald chose to interpret them as evidence of the bankruptcy of materialism and of the need for a methodological unification around the concept of energy. In the last analysis, Ostwald was motivated by a generous optimism and was convinced of making a positive contribution toward a science of nature free from hypotheses, that is, free from assumptions or "metaphysical" entities such as atoms and molecules. Kelvin, instead, having recognized the failure of a research project on which he had toiled for more than half a century, chose not to look for a philosophical way out.

It is instructive to compare the different arguments Ostwald and Kelvin used to deal with two problems they necessarily shared: electromagnetism and statistical mechanics.

At the 1895 meeting in Lübeck, during which the majority of German scientists violently attacked the fundamental themes of Boltzmann's thought, Wilhelm Ostwald delivered a lecture entitled "Beyond Scientific Materialism."[21] It would serve no purpose here to discuss the opinions expressed at that meeting, its philosophical objectives, or the answers Ostwald received from various quarters. What is relevant to our discussion is a particular statement made by Ostwald to the effect that the mechanistic view of the world is untenable for reasons that are strictly physical. "The proposition that all natural phenomena can be reduced to mechanical ones cannot even be taken as a useful working hypothesis," Ostwald writes. "It is simply a mistake. This mistake is clearly revealed by the following fact. All the equations of mechanics have the property that they admit of sign inversion in the temporal quantities. That is to say, theoretically perfect mechanical processes can develop equally well forward or backward in time. Thus, in a purely mechanical world there could not be a before or an after as we have in our world: the plant could become seed again, the butterfly turn back into caterpillar and the old man into child. No explanation is given by the mechanistic doctrine for the fact that this does not happen, nor can it be given because of the fundamental property of mechanical equations. The *de facto* irreversibility of natural phenomena thus proves that there are processes that cannot be described by mechanical equations; and with this the verdict on scientific materialism is settled."

This statement contains at least three theses. The first lumps together most nineteenth-century research in mathematical physics under the heading of mechanistic explanation. The second maintains that the mechanistic explanation is contradictory in the question of irreversibility because irreversibility cannot be accounted for by equations that admit of time reversal. The third thesis equates the failure of the mechanistic explanation with the failure of scientific materialism.

The first thesis evidences an exceedingly superficial view of the historical process in question. The fact that Ostwald's thesis had a large following is completely irrelevant if it is true that in matters of physics problems are not solved by popular vote. As for the second thesis, Ostwald's version of the paradox of time and velocity reversal in mechanical systems is a poor rehashing of the controversy that was set off by Kelvin[22] in 1874 and grew particularly lively in the two following decades at the hands of Boltzmann's critics. When viewed solely in the context of rational mechanics, the paradox of reversal is not without ambiguities; it becomes a serious problem when it involves the relationship between rational mechanics and the concept of probability. The fact that Ostwald should present it as a paradox arising from a direct confrontation between the mechanistic explanation and irreversible natural processes is something that concerns Ostwald alone, although it shows that the faith in energetics often rests on very shaky bases. The third thesis is based on the stratagem of equating the philosophical idea of matter with the idea of matter operating in the physical world, as though it were the physicists who ought to prove that the philosophers' matter exists, whereas it is really the philosophers who should modify their ideas according to the physicists' findings.

Ostwald's argument against materialism in general is faulty not because it is directed against materialism, but because it draws improper conclusions from three basically unsound theses.

Nor does Ostwald muster more convincing arguments against the materialistic mechanism that in his opinion many physicists have attempted to introduce into the foundations of the electromagnetic theory of light or of the wave theory viewed as a chapter in mechanics. Having considered only the theoretical difficulties involved in the concept of ether, Ostwald, a chemist, concludes from them that the ether does not exist as a "physical" object. Such a conclusion is not serious, based as it is on Hertz's remark that Maxwell's theory is nothing more than Maxwell's equations. The latter is of course an interesting point of view on Maxwell's equations, but certainly offers no solution for the problem of the ether. It is a curious thing that when Hertz's celebrated remark is quoted, no mention is made of the fact that he was an intelligent advocate of research into the physics of hidden structures.

Let us now turn from the philosophical conference of 1895 to Kelvin's lecture of 1900 on "Nineteenth Century Clouds over the Dynamical Theory of Heat and Light."[23]

The basic difference between Ostwald's and Kelvin's views can be reduced in the last analysis to a difference between the idea of crisis and that of difficulty. According to Ostwald, the development of the physical sciences has reached a crisis that can only be solved by a philosophical verdict on the disappearance of matter, and by the proposition that the predicate of reality should be given to energy. According to Kelvin, instead, the development of the physical sciences has met with very serious difficulties, which, however, can be overcome by reformulating the unresolved problems with new mathematical physical concepts.

Kelvin's clouds consist of two fundamental theoretical knots. The first involves an unanswered question about the undulatory theories accepted at the beginning of the century: "How could the earth move through an elastic solid such as essentially the luminiferous ether?" The seriousness of this difficulty, Kelvin writes, stems from the fact that there are no inconsistencies in the explanation of the ordinary phenomena of terrestrial optics, while an inconsistency is evident as soon as we discuss the "conclusion that ether in the earth's atmosphere is motionless relatively to the earth" with reference to the Michelson–Morley experiment and the "brilliant suggestion" on contraction by Fitzgerald and Lorentz.

This thicket of difficulties is what Kelvin calls "cloud No. 1." The second cloud appears when the Boltzmann–Maxwell doctrine on the partition of energy is defined in terms of a theorem. On the strength of the subtle analysis performed in the last decade of the century not only by Kelvin himself but by Boltzmann, Poincaré, and Rayleigh, the lecture of 1900 puts strong emphasis on a twofold argument: "It is not quite possible to rest contented with the mathematical verdict not proved, and the experimental verdict not true, in respect to the Boltzmann–Maxwell doctrine."

Now, if what Boltzmann and the other statisticians are talking about is a theorem, then there must be a demonstration somewhere. But Boltzmann, Kelvin writes, has never demonstrated anything in this respect. As for Maxwell and his generalization of Boltzmann's theorem for systems with any number of degrees of freedom, Kelvin adds, we can only repeat that it is a magnificent generalization of a principle rather than a generalization of a theorem that has been proved true.[24]

And even if a demonstration were to exist somewhere, we could at best build on it a theory invalidated by experimental data. If we indicate by k the relationship between thermal capacity at constant pressure and thermal capacity at constant volume, Kelvin writes, the following table clearly shows the theory's position with respect to the experimental data.

"This notable divergence from observation," Kelvin writes, "suffices to disprove absolutely the Boltzmann–Maxwell doctrine." But this divergence does not adequately reflect the falseness of the theory: the situation

| Gas | Values of $k - 1$ | |
	According to the Boltzmann–Maxwell doctrine	By observation
Air	$2/7 = 0.2857$	0.406
H_2	" "	0.40
O_2	" "	0.41
Cl_2	" "	0.32
CO	" "	0.39
NO	" "	0.39
CO_2	$1/6 = 0.1667$	0.30
N_2O	" "	0.331
NH_3	$1/9 = 0.1111$	0.311

becomes even more serious when we take into account the data relative to molecular spectra in gases, from which we find that the number of degrees of freedom for each molecule is "enormously larger" than that used in the determination of k. "There is in fact no possibility of reconciling the Boltzmann–Maxwell doctrine with the truth regarding the specific heats of gases."

Cloud No. 2 is so serious that even Rayleigh, who defends many aspects of Boltzmann's theory, has to recognize the "destructive simplicity" of its conclusions. Kelvin agrees with Rayleigh's statement that "we are brought face to face with a fundamental difficulty, relating not to the theory of gases merely, but rather to general dynamics"; but contrary to Rayleigh's opinion he suggests that the removal of the cloud entails the elimination from Boltzmann's theory of its conclusion as to the principle of the partition of energy.

In a short paper published in 1900 in the *Philosophical Magazine*[25] Rayleigh reaches a peculiar a priori result from an analysis of Cloud No 2. Blackbody radiation is a function of the absolute temperature θ and the wavelength λ. On the strength of "arguments [of] considerable weight" presented by Boltzmann and Wien, it must be concluded that the function assumes the form

$\theta^5 \Phi(\theta\lambda) \, d\lambda.$

A "further specialization" of this form led Wien[26] to write that the energy of that part of the spectrum comprised between λ and $\lambda + d\lambda$ is given by

$c_1\lambda^{-5}e^{-c_2/(\lambda\theta)} \, d\lambda.$ \hfill (A)

In Rayleigh's opinion, (A) is something "strange." Judged from the theoretical point of view, (A) "appears to me to be little more than a conjecture," although Planck has furnished some support for it in the context of thermodynamics and Paschen has confirmed it experimentally. (A) is indeed difficult to accept for it has peculiar implications, notably the prediction that for an increasing θ the radiation of a given λ tends to a limit.

The actual problem, however, is not easy to formulate. According to Rayleigh, "speculation about this subject is hampered by the difficulties which attend the Boltzmann–Maxwell doctrine of the partition of energy," although these difficulties do not prevent the formulation of a notable variant of (A):

$$c_1\theta\lambda^{-4}e^{-c_2/(\lambda\theta)}\,d\lambda, \qquad \text{or} \qquad c_1\theta k^2 e^{-c_2 k/\theta}\,dk, \tag{B}$$

where $k^2 = p^2 + q^2 + r^2$. The p, q, and r parameters may be interpreted as the coordinates of points forming a cubic reticule, with k as the distance of a point from the origin of the axes. The number of points for which k falls between k and $k + dk$ is given by $k^2\,dk$, and "this expresses the distribution of energy according to the Boltzmann–Maxwell law."

Rayleigh admits he is unable to choose between (A) and (B), although (B) appears to be more satisfactory for large values of $\lambda\theta$. The choice, in his opinion, should be made by distinguished experimenters. In a brief final note added in 1902 Rayleigh remarks that the prediction for very high wavelengths has been verified by the experiments of Rubens and Kurlbaum and that "the formula of Planck, given about the same time, seems best to meet the observations."[27]

An examination of what may be called the dynamics of the problem embodied in Cloud No. 2 clearly reveals the contrast between the futility of Ostwald's proposal and the lucid analysis contained in Kelvin's criticism.

Heavy clouds are certainly hanging over the sciences at the turn of the century, but they do not portend a crisis to be resolved by a verdict handed down by philosophical tribunals. Although a philosophical tribunal may suggest a particular attitude in the investigation of objects and theories, it is wholly incompetent to direct scientific research. The latter is rigorously based on areas of rules and it is within this system of reference that problems, difficulties, paradoxes, and solutions arise.

When we argue that problems, difficulties, paradoxes, and solutions have the attribute of objectivity, we are not making a careless concession to some sort of naive scientism or outdated materialism. Rather, we wish to point out that such problems, difficulties, paradoxes, and solutions are neither the products of intuitive flashes or personal faith, of a scientist's psychological upheaval to be described in terms of Gestalt theory, nor the products of an ideological invasion that muddles the precepts of scientific inquiry to pay homage to the cultural dominance of a particular social class. In other words, we wish to refute the thesis, upheld by Michael Polanyi for instance, that a scientific problem does not exist unless it is of interest to somebody.[28] With such statements Polanyi prepares the philosophical ground for the burial of the natural sciences. Reducing objective problems and their dynamics to purely personal facts inevitably leads to a cognitive catastrophe that, in essence, consists of an old and worn-out philosophical conceit, namely, that the history of science is a series of routine exercises interrupted here and there by the magical flashes of some solitary genius. Romanticism, like all other philosophies, always survives in its worst aspects.

Notes and Bibliography

Part I

REFLECTIONS ON THE HISTORY OF THE PHYSICAL SCIENCES

Chapter 1 The Scientist's Dictionary

1. Mark Twain's saying of course lends itself to different interpretations. See G. Temple, "From the Relative to the Absolute," in *Turning Points in Physics*, Amsterdam, 1959; G. Toraldo di Francia, *L'Indagine del Mondo Fisico*, p. 16, Torino, 1976.

2. M. Faraday, "Thoughts on Ray-Vibrations," *Phil. Mag.* 24 (1846), 136; idem, "On the Possible Relation of Gravity to Electricity," *Phil. Trans.* 1 (1851); idem, "On the Conservation of Force," *Proc. Royal Soc.* 2 (1857), 352.

3. J. C. Maxwell, "Note on the Attraction of Gravity: A Dynamical Theory of the Electromagnetic Field," *Phil. Trans. Royal Soc.* 155 (1865), 459.

4. J. C. Maxwell, "On Boltzmann's Theorem on the Average Distribution of Energy in a System of Material Points," *Cambridge Phil. Trans.* 3 (1879), 547.

5. J. C. Maxwell, "A Discourse on Molecules," *British Assoc. Bradford Phil. Mag.* 46 (1873), 453.

Chapter 2 The Galileo–Dirac Proposition

1. P. Duhem, Σώζειν τά φαινόμενα. *Essai sur la notion de théorie physique de Platon à Galilée*, Paris, 1908.

2. P. Feyerabend, "Problems of Empiricism," in *Beyond the Edge of Certainty*, edited by Colodny, 1965.

3. P. Feyerabend, "Consolations for the Specialist," in *Criticism and the Growth of Knowledge*, edited by I. Lakatos and A. Musgrave, Cambridge University Press, London, 1979.

4. See chapter 3.

5. I. Lakatos, "Falsification and the Methodology of Scientific Research Programmes," in *Criticism and the Growth of Knowledge*.

6. Ibid.

7. L. Geymonat, *Galileo Galilei*, Torino, 1957.

8. Galileo, *Dialogo sopra i due massimi sistemi del mondo*, Second Day, VII, pp. 233–234, 1632.

9. Ibid., p. 139. The liberalization of Galilean thought requires that the celebrated passage from the *Essayer* that describes the book of nature as "written in mathematical characters" and "open to our eyes" should be interpreted in such a way as to remove the temptation of absolute knowledge, and to emphasize the need for a distinction between those books "in which the least important thing is that what is written in them should be true" and mathematical reasoning: without the latter, as Galileo remarks, human thought becomes "an aimless wandering through a dark maze."

10. B. Riemann, *Gesammelte mathematische Werke und wissenschaftlicher Nachlass*, edited by H. Weber, Leipzig, 1876 (posthumous ed.). Reprinted as *Collected Papers*, New York, 1953.

11. See chapter 4.

12. P. A. M. Dirac, "Quantitized Singularities in the Electromagnetic Field," *Proc. Royal Soc. (A)* 133 (1931).

Part II

STUDIES ON THE SECOND SCIENTIFIC REVOLUTION

Chapter 3   Herschel's Lion

1. P. G. Tait, "On the Foundations of the Kinetic Theory of Gases," *Phil. Mag.* 21 (1886); idem, "On the Foundations of the Kinetic Theory of Gases," *Phil. Mag.* 23 (1887); L. Boltzmann, "On the Assumptions Necessary for the Proof of Avogadro's Law," *Phil. Mag.* 23 (1887) (transl. from *Sitz. König. Akad. Wien* 94 (1887)); P. G. Tait, "The Assumptions Required for the Proof of Avogadro's Law," *Phil. Mag.* 23 (1887); L. Boltzmann, "On Some Questions in the Kinetic Theory of Gases," *Phil. Mag.* 25 (1888) (transl. from *Wied. Ann.* 96 (1888)); S. H. Burbury, "On the Diffusion of Gases; A Reply to Prof. Tait," *Phil. Mag.* 25 (1888); P. G. Tait, "On Some Questions in the Kinetic Theory of Gases: A Reply to Prof. Boltzmann," *Phil. Mag.* 25 (1888).

2. The theme of "just a piece of mathematics" is one that recurs often in the fight against Boltzmann's physics. Particularly important in this respect is the polemic conducted in the magazine *Nature* from 1894 to 1895, which hinges on the question whether the H theorem is nothing but a naked theorem without any empirical confirmation. This polemic originates in 1890 with Culverwell's analysis of the relation between

irreversibility and dynamics: E. P. Culverwell, "Note on Boltzmann's Kinetic Theory of Gases, and on Sir W. Thomson's Address to Section A, Brit. Ass., 1884," *Phil. Mag.* 30 (1890); W. Strutt (Lord Rayleigh), "Remarks on Maxwell's Investigations regarding Boltzmann's Theorem," *Phil. Mag.* 33 (1892); G. H. Brian, "On the Present State of our Knowledge of Thermodynamics," *British Assoc. Rep.* (1891). On the related question of the equipartition of energy see: W. Thomson (Lord Kelvin), "On a Decisive Test-Case Disproving the Maxwell–Boltzmann Doctrine regarding Distribution of Kinetic Energy," *Proc. Royal Soc.* 51 (1892), and *Phil. Mag.* 33 (1892). In particular, see: E. P. Culverwell, "Dr. Watson's Proof of Boltzmann's Theorem on Permanence of Distribution," *Nature*, 25 October 1894; S. H. Burbury, "Boltzmann's Minimum Function," *Nature*, 22 November 1894; E. P. Culverwell, "The Kinetic Theory of Gases," *Nature*, 22 November 1894; H. W. Watson, "Boltzmann's Minimum Theorem," *Nature*, 29 November 1894; E. P. Culverwell, "Letter," *Nature*, 29 November 1894; S. H. Burbury, "The Kinetic Theory of Gases," *Nature*, 20 December 1894; G. H. Brian, "The Kinetic Theory of Gases," *Nature*, 31 January 1895; S. H. Burbury, "Boltzmann's Minimum Function," *Nature*, 31 January 1895; L. Boltzmann, "Letter," *Nature*, 18 April 1895; E. P. Culverwell, "Prof. Boltzmann's Letter on the Kinetic Theory of Gases," *Nature*, 18 April 1895; G. H. Brian, "The Assumptions in Boltzmann's Minimum Theorem," *Nature*, 9 May 1895; E. P. Culverwell, "Boltzmann's Minimum Function," *Nature*, 13 June 1895; L. Boltzmann, "On the Minimum Theorem in the Theory of Gases," *Nature*, 4 July 1895; S. H. Burbury, "The Kinetic Theory of Gases," *Nature*, 4 July 1895. The best source for a subtle analysis of these problems is still the 1911 classical monograph by P. and T. Ehrenfest, *Begriffliche Grundlagen der statistichen Auffassung in der Mechanik.* Encyk. Math. Wiss., vol. 4, Leipzig and Berlin, 1911 (English transl. Ithaca, 1959). Worthy of note is Brush's historical reconstruction of various aspects of the problem under discussion. In particular, see: S. G. Brush, "Interatomic Forces and Gas Theory from Newton to Lenard–Jones," *Arch. Rational Mech. and Anal.* 39 (1970), 1; idem, "The Development of the Kinetic Theory of Gases, vii: Randomness and Irreversibility," *Arch. Hist. Exact Sci.* 12 (1974), 1.

3. P. G. Tait, *Lecture on Some Recent Advances in Physical Science*, London, 1876.

4. Ibid., p. 25.

5. W. Thomson (Lord Kelvin) and P. G. Tait, *A Treatise on Natural Philosophy*, 2 vols., London, 1879.

6. J. B. Fourier, *Théorie analytique de la chaleur,* Paris, 1822 (see J. B. Fourier, *Oeuvres,* edited by G. Darboux, Paris, 1890). Among the recent works on Fourier, two are particularly interesting: I. Grattan-Guinness and J. R. Ravetz, *Joseph Fourier, 1768–1830,* Cambridge, Mass., 1972; J. Herivel, *Joseph Fourier, The Man and the Physicist,* Oxford, 1975. These two monographs may be considered complementary. However, their authors express different opinions about Fourier's position with regard to Laplace's physics. The former work emphasizes the radical character of Fourier's critique in the question of the relationship between mathematical and physical research; in the latter, instead, Fourier's thesis that the theory of heat cannot be reduced to the principles of mechanics is judged as a *curious assertion,* a *puzzling statement.* According to Herivel, Fourier's position is explainable in part "at the level of the subconscious," as a "reluctance to see his own theory taken under Laplace's newtonian umbrella" (pp. 226–227, 234).

7. J. B. Fourier, *Théorie analytique de la chaleur,* p. 17, Breslau, 1883.

8. Ibid., p. 584.

9. Ibid., pp. 191–193.

10. Ibid., p. 584.

11. E. Bellone, "Il Significato metodologico dell'eliminazione dei modelli di calorico promossa da J. Fourier," *Physis* 9 (1967).

12. J. B. Fourier, *Théorie analytique de la chaleur,* II–III, Breslau, 1883.

13. Ibid., pp. 13–14.

14. Ibid., p. 18.

15. Ibid., p. 585.

16. P. Casini, *L'Universo-macchina. Origini della filosofia newtoniana.* Bari, 1969; idem, *Introduzione all'illuminismo. Da Newton a Rousseau,* Bari, 1973; A Rupert Hall, *From Galileo to Newton, 1630–1720,* London, 1963.

17. See chapter 7.

18. R. Harris Inglis, "Address," in *Report of the Seventeenth Meeting of the British Association for the Advancement of Science; Held at Oxford in June 1847,* London, 1848.

19. W. Thomson (Lord Kelvin) and P. G. Tait, *A Treatise on Natural Philosophy,* vol. 1, pp. 219–220, London, 1879.

20. Ibid., p. 445.

21. Ibid., vol. 2, pp. 3–4.

22. Ibid., vol. 1, p. 219.

23. Ibid., p. 441.

24. As we know, the idea that mathematics is the servant of physics is inherent in the idea that theoretical physicists are translators.

25. P. G. Tait, *Lectures on Some Recent Advances in Physical Science*, pp. 283 and foll.

26. E. Du Bois-Reymond, *Über die Grenzen des Naturerkennens*, 3rd ed., Leipzig, 1891.

27. E. Du Bois-Reymond, "Die sieben Welträthsel."

28. P. G. Tait, *Lectures on Some Recent Advances in Physical Science*, p. 4.

29. Ibid., pp. 5–6.

30. B. Riemann, *Über die Hypothesen welche der Geometrie zu Grunde liegen* (1854) (see *Collected Papers*).

31. Ibid.

32. P. G. Tait and W. J. Steele, *A Treatise on the Dynamics of a Particle*, 2nd ed., London, 1865.

33. P. G. Tait, "On the Size of Atoms," *Nature*, March 1870.

34. W. Thomson (Lord Kelvin), *Baltimore Lectures on Molecular Dynamics and the Wave Theory of Light*, Cambridge, 1884.

35. W. Thomson (Lord Kelvin), "On a Decisive Test-Case Disproving the Maxwell–Boltzmann Doctrine regarding Distribution of Kinetic Energy," *Proc. Royal Soc.* 51 (1892).

36. W. Thomson (Lord Kelvin), "The Dynamical Theory of Heat," *Trans. Royal Soc. Edinburgh*, March 1851; idem, *Phil. Mag.* 4 (1852).

37. For the relation between Kelvin, Joule, and Carnot, see chapter 7.

38. W. Thomson (Lord Kelvin), "On a Universal Tendency in Nature to the Dissipation of Mechanical Energy," *Phil. Mag.* 4 (1852). See also Kelvin's "On the Mechanical Action of Heat and Light; On the Power of Animated Creatures over Matter; On the Sources Available to Man for the Production of Mechanical Effects," ibid., 256.

39. The view of knowledge as a succession of increasingly better approximations implies a cumulative process if its development is not assumed to occur by reinterpretations. Kelvin and Tait believe that for science to grow on sound bases some general principles discovered by induction must remain unaltered.

40. This is a truly classical problem in that it revolves on the theme of the arrow of time. See, for example, Sun-Tak Hwang, "A New Interpretation of Time Reversal," *Foundations Phys.* 2 (1972), 4.

41. E. Bellone, *Opere di Kelvin*, vol. 34, Torino, 1971. See also J. Larmor, "Obituary Notice of W. Thomson, Baron of Kelvin," *Proc. Royal Soc.* 81 (1908), iii–lxxxvi.

42. W. Thomson (Lord Kelvin), "On Vortex-Atoms," *Phil. Mag.* 34 (1867).

43. R. Clausius, "On the Second Fundamental Theorem of the Mechanical Theory of Heat: Address to the Forty-First Meeting of German Scientists Held at Frankfurt," *Phil. Mag.* 35 (1868).

44. See Kelvin's thesis on the age of the sun's heat, *Popular Lectures and Addresses,* I–III, London, 1889–1894.

45. W. Thomson (Lord Kelvin), "Address, ciii" (1871), in *Report of the Forty-First Meeting of the British Association for the Advancement of Science; Held at Edinburgh in August 1871,* London, 1872.

46. G. Lamé, *Cours de physique de l'École Polytechnique,* Brussels, 1836.

47. Quite interesting in this respect are Kelvin's youthful writings published from 1841 in the *Cambridge Mathematical Journal.* See W. Thomson (Lord Kelvin), *Mathematical and Physical Papers,* vol. 1, edited by Lord Kelvin, 1882. This volume comprises essays and papers written from 1841 to 1853. Some of Kelvin's comments on his earlier writings help to document and clarify the relationship between the young mathematical physicist and the French school.

48. G. Lamé, *Leçons sur la théorie analytique de la chaleur,* p. xi, Paris, 1850.

49. S. D. Poisson, *Théorie mathématique de la chaleur,* Paris, 1835.

50. For a particularly acute analysis of William Hamilton's position see R. Olson, *Scottish Philosophy and British Physics, 1750–1880,* Princeton, 1975. Without entering into the details of the "Kelvin problem," Olson proposes a partial revision of Duhem's and Poincaré's interpretation with regard to the British scientific style.

51. Ibid., p. 70.

52. W. Thomson (Lord Kelvin), "On a Mechanical Representation of Electric, Magnetic and Galvanic Forces," *Cambridge and Dublin Math. J.* 2(1847) (*Mathematical and Physical Papers,* vol. 1, p. xxvii). For the significance of this brief paper see J. Larmor, "Obituary Notice of William Thomson, Baron Kelvin of Largs, 1824–1907," *Proc. Royal Soc.* 81 (1908).

53. M. Faraday, "Thoughts on Ray-vibrations," *Phil. Mag.* 28 (1846). It is a letter to Richard Phillips, reprinted in 1859 in *Experimental Researches in Chemistry and Physics.*

54. Kelvin will later emphasize the role played by Joule's ideas in this revision.

55. Kelvin and Tait were certainly influenced by Helmholtz's improved formulation of the theory of differential equations for hydrodynamics and by Riemann's studies in multiple continuity, as well as by Rankine's

investigations of hydrodynamic models: the genesis of the concept of vortex-atom is incomprehensible without reference to these advances in mathematics and mathematical physics.

56. Tait's famous experiment on the collisions of smoke rings is effectively described by Tait himself in P. G. Tait, *Lectures on Some Recent Advances in Physical Science.*

57. W. Thomson (Lord Kelvin), "On Vortex-Atoms."

58. Ibid., p. 4.

59. W. Thomson (Lord Kelvin), "The Six Gateways of Knowledge," address of 3 October 1883 at the Birmingham and Midland Institute, in *Popular Lectures and Addresses,* vol. 1.

60. W. Thomson (Lord Kelvin), *Baltimore Lectures.*

61. W. Thomson (Lord Kelvin), "On Vortex-Atoms."

62. It is not a matter of taste. Such romantic ambiguities result from attempts to subject a mathematical physical theory to analysis by means of philosophical concepts. See the excerpt quoted in note 22, chapter 7.

63. W. Thomson (Lord Kelvin), *Baltimore Lectures.* Editing the text for publication in 1904, and in an article written for the 27 May 1897 issue of *Nature,* Kelvin recognized the essential validity of the theories of the discontinuum and atomic structure, referring to Varley, Crookes, Faraday, Maxwell, and Helmholtz and suggesting that Aepinus' work should be reinterpreted with the theoretical tools provided by Boscovich.

64. W. Thomson (Lord Kelvin), "On Vortex-Atoms."

65. H. Poincaré, *Théorie des tourbillons. Leçons professées pendant le deuxième semestre 1891–92,* edited by A. Lamotte, Paris, 1893.

66. P. G. Tait, *An Elementary Treatise on Quaternions,* Oxford, 1867.

67. W. R. Hamilton, "On a General Method of Expressing the Paths of Light, and of the Planets, by the Coefficients of a Characteristic Function." *Dublin Univ. Rev.* (1833).

68. W. R. Hamilton, "Theory of Conjugate Functions, or Algebraic Couples; with a Preliminary and Elementary Essay on Algebra as the Science of Pure Time," *Trans. Royal Irish Acad.* 17 (1837) (see *The Mathematical Papers of Sir William Rowan Hamilton,* vol. 3, edited by H. Halberstam and R. E. Ingram, Cambridge, 1967).

69. W. R. Hamilton, "Preface to Lectures on Quaternions," in *The Mathematical Papers of Sir William Rowan Hamilton.*

70. P. G. Tait, *An Elementary Treatise on Quaternions,* chapter 11, "Physical Applications."

71. Ibid., p. ix.

72. See "Introduction," *The Mathematical Papers of Sir William Rowan Hamilton*.

73. P. G. Tait, *An Elementary Treatise on Quaternions*, p. 4.

74. Ibid., p. v.

75. Ibid., p. vi.

76. Ibid., pp. vi–vii.

77. Ibid., p. 4.

78. J. F. W. Herschel, *Preliminary Discourse on the Study of Natural Philosophy*, London, 1830. A facsimile edition, edited by M. Partridge, was published in 1966, New York.

79. Ibid., p. 50.

80. Ibid., p. 40.

81. Ibid., p. 108.

82. Ibid., part II, chapter 1: "Of Experience as the Source of our Knowledge—Of the Dismissal of Prejudices—Of the Evidence of our Senses."

83. Ibid., pp. 6–21.

84. Ibid., pp. 29–33.

85. Ibid., p. 104.

86. Ibid., part II, chapter 1.

87. Ibid., p. 76.

88. Ibid., p. 78.

89. Ibid., p. 80.

90. Ibid., p. 84.

91. Ibid., p. 86. See note 100.

92. Ibid., pp. 96–97.

93. Ibid., pp. 113–116.

94. Ibid., p. 186.

95. Ibid., pp. 150–151.

96. Ibid., pp. 164–166.

97. Ibid., p. 178.

98. Ibid., pp. 174–175.

99. See the discussion in chapter 7.

100. J. F. W. Herschel, *Preliminary Discourse*, p. 179.

101. Ibid., p. 178. It should be kept in mind that for Herschel the laws of mechanics are rigorous experimental laws. On the one hand, Herschel is aware of a *degree of obscurity* at the basis of the theory of motion: "How far we may ever be enabled to attain a knowledge of the ultimate and inward processes of nature in the production of phenomena, we have no means of

knowing; but, to judge from the degree of obscurity which hangs about the only case in which we feel within ourselves a *direct* power to produce any one, there seems no great hope of penetrating so far. The case alluded to is the production of motion by the exertion of force" (p. 86). On the other hand, the science of motion is the best experimental science available: "By far the most general phenomenon with which we are acquainted, and that which occurs most constantly, in every inquiry into which we enter, is motion and its communication. Dynamics, then, or the science of force and motion, is thus placed at the head of all the sciences; and, happily for human knowledge, it is one in which the highest certainty is attainable, a certainty no way inferior to mathematical demonstration. As its axioms are few, simple, and in the highest degree distinct and definite, so they have at the same time an immediate relation to geometrical quantity, space, time and direction, and thus accommodate themselves with re-markable facility to geometrical reasoning. Accordingly, their conse-quences may be pursued, by arguments purely mathematical to any extent, insomuch that the limit of our knowledge of dynamics is de-termined only by that of pure mathematics, which is the case in no other branch of physical science" (p. 96). Herschel's definition of law of nature plays a fundamental role in his view of science: "A statement in words of what will happen in such and such proposed general contingencies" (90). Thus his use of the term axiom must not be misunderstood. Herschel writes, for instance, "The law of gravitation is a physical axiom of a very high and universal kind, and has been raised by a succession of inductions and abstractions drawn from observation of numerous facts and sub-ordinate laws in the planetary system" (p. 98). From this point of view a physical axiom is a basis for reasoning, in the sense that it enables mathematics to deduce conclusions that are "true in fact."

102. Ibid., p. 195.
103. Ibid., p. 197.
104. Ibid., p. 204.
105. Ibid., p. 206.
106. Ibid., p. 219.
107. Ibid., p. 264.
108. Ibid., p. 348.
109. P. Duhem, *La théorie physique: Son objet, sa structure*, Paris, 1906.
110. L. Geymonat, *Storia del pensiero filosofico e scientifico*, vol. 3, p. 153, Milano, 1971.
111. Quoted from R. Olson, *Scottish Philosophy and British Physics, 1750–1880*, pp. 29–31.

112. Ibid., p. 31.

113. Ibid., p. 73.

114. W. Thomson (Lord Kelvin), "The Six Gateways of Knowledge."

115. See Olson, *Scottish Philosophy and British Physics, 1750–1880.*

116. L. Boltzmann, "Über die Grundprinzipien und Grundgleichungen der Mechanik," lecture at Clark University, in *Populäre Schriften,* pp. 253–307, 1905. L. Boltzmann, "Über die Bedeutung von Theorien," lecture at Graz, in *Populäre Schriften,* pp. 76–80. For a very good selection of Boltzmann's writings in English translation see B. McGuinness, *L. Boltzmann. Theoretical and Philosophical Problems. With an Introduction by S. R. De Groot,* Dordrecht and Boston, 1974.

Chapter 4   In Praise of Theory

1. L. Boltzmann, "Über die Bedeutung von Theorien," in *Populäre Schriften,* Leipzig, 1905. English transl. B. McGuinness, *L. Boltzmann. Theoretical and Philosophical Problems,* Dordrecht and Boston, 1974, pp. 33–36. From now on we will refer to this latter edition as LB and to the 1905 German edition as PS.

2. L. Boltzmann, "Der zweite Hauptsatz der mechanischen Wärmetheorie," (1886), PS, pp. 25–50; LB, pp. 13–32.

3. L. Boltzmann, "Über die unentbehrlichkeit der Atomistik in der Naturwissenschaft" (1897), PS, pp. 141–157; LB, pp. 41–53.

4. Ibid., LB note 8.

5. L. Boltzmann, "Über die mechanische Bedeutung des zweiten Hauptsatzes der Wärmetheorie," *Sitz. König. Akad. Wien* 53 (1866).

6. L. Boltzmann, "Über des Wärmegleichgewicht zwischen mehratomigen Gasmolekülen," ibid. 63 (1871); "Einige allgemeine Sätze über Wärmegleichgewicht," ibid. 63 (1871); "Analytischer Beweis des zweiten Hauptsatzes der mechanischen Wärmetheorie aus den Sätzen über das Gleichgewicht der lebendingen Kraft," ibid. 63 (1871).

7. L. Boltzmann, *Weitere Studien über das Wärmegleichgewicht unter Gasmolekülen, Sitz. König. Akad. Wien* 66 (1872).

8. Ibid., p. 307.

9. L. Boltzmann, "Über die Beziehung zwischen dem zweiten Hauptsatze der mechanischen Wärmetheorie und der Wahrscheinlichkeitsrechnung respektive den Sätzen über das Wärmegleichgewicht," ibid. 76 (1877).

10. For Maxwell's position see his "A Discourse on Molecules," (footnote 5, chapter 1) and his treatise *Theory of Heat,* London, 1871. Also interesting

is *Science and Free Will,* quoted in *The Life of J. C. Maxwell,* by L. Campbell and W. Garnett, London, 1882–84, reprinted in 1969 and edited by R. H. Kargon.

11. L. Boltzmann, "Weitere Studien über das Wärmegleichgewicht unter Gasmolekülen."

12. L. Boltzmann, "Über die Frage nach der objektiven Existenz der Vorgänge in der unbelebten Natur" (1897), PS, pp. 162–187; LB, pp. 57–76.

13. Ibid., LB, notes 12 and 14.

14. LB, note 1.

15. L. Boltzmann, "Über die Entwicklung der Methoden der theoretischen Physick in neuerer Zeit" (1899), PS, pp. 198–227; LB, pp. 77–100.

16. L. Boltzmann, "Über die Grundprinzipien und Grundgleichungen der Mechanik" (1899), PS, pp. 253–307; LB, pp. 101–128.

17. M. Planck, "Gegen die neuere Energetik," *Wied. Ann.* 57 (1896), 72.

18. L. Boltzmann, "Über die Prinzipien der Mechanik" (1900–1902), PS, pp. 308–337; LB, pp. 129–152.

19. Boltzmann's position is summarized in the paragraph "On What Exists" in this same chapter.

20. L. Boltzmann, "Ein Antrittsvortrag zur Naturphilosophie" (1903), PS, pp. 338–344; LB, pp. 153–158.

21. L. Boltzmann, "Über statistische Mechanik" (1904), PS, pp. 345–363; LB, pp. 159–172.

22. L. Boltzmann, "Über eine These Schopenhauers" (1905), PS, pp. 385–402; LB, pp. 185–198.

23. L. Boltzmann, "Vorlesungen über die Principe der Mechanik" (1904), part II, LB, p. 257.

24. W. Thomson (Lord Kelvin), "Kinetic Theory of the Dissipation of Energy," *Proc. Royal Soc. Edinburgh,* February 1874; published in *Nature,* 9 April 1874; reprinted a few years later during the controversy with Boltzmann in *Phil. Mag.* 33 (1892).

25. J. Loschmidt, "Über den Zustand des Wärmegleichgewichtes eines Systems von Körpern mit Rücksicht auf die Schwerkraft (I–IV)," *Sitz. König. Akad. Wien* 73 (1876), 135; 73 (1876), 336; 75 (1877), 67; 76 (1877), 209.

26. Y. Elkana, "Boltzmann's Scientific Research Programme and its Alternative," in *Some Aspects of the Interaction between Science and Philosophy.*

Proc. Symp. Van Leer Jerusalem Foundation, January 1971. Elkana's work is fundamental in showing the need for reinterpreting the interaction between philosophical norms and mathematical-physical problems in Boltzmann's thought. According to Elkana, however, the first shift in Boltzmann's program occurred after Loschmidt's criticism. A second shift would have occurred in the 1890s, as a result of the Boltzmann–Ostwald–Mach debate and the analyses of the Zermelo–Poincaré paradox. According to the author, Boltzmann's program turns on the proposition that usefulness, not truth, is the basic requirement of method and hypotheses. Quite rightly, Elkana disagrees with any attempt to portray Boltzmann as a precursor of neopositivism, pointing out that Boltzmann never renounced the explicative power of science.

27. To the naive Marxist: with this observation we do not automatically defend internal history after criticizing it. We only wish to point out (in the text) that research in theoretical physics can enjoy a large measure of autonomy. We will remind him that in science autonomy and philosophical involvement are not mutually exclusive. By "naive Marxists" we mean those who believe in the duty—or at least the propriety—of sustaining the following opinions: (a) science can be reduced to ideological forms; (b) historical materialism is the basis whereby external history is the only history of science. Astonishing conclusions can be drawn from (a) and (b): for instance, non-Euclidean geometries become weapons in a clever anti-working class maneuver, and the 1929 depression is held as the decisive factor in the failure (?) of Hilbert's mathematical studies.

28. L. Geymonat, *Filosofia e filosofia della scienza*, 5th ed., Milano, 1970.

29. "Turning point" may be used to mean a revolution; but also a revolution is a historical process, not a gap in history or a sudden jump.

30. One gets the impression that the conflict between the advocates of continuity and the advocates of discontinuity is due to a quantitative difference in the number of documents they consult, coupled with a difference of opinion on the procedure for establishing a credible temporal sequence. In both cases, however, the history of science is forced into a straitjacket.

31. Interesting reflections on the question of "concepts in flux" are contained in Y. Elkana's "Helmholtz' 'Kraft': An Illustration of Concepts in Flux," *Hist. Studies Phil. Sci.* 2 (1970).

Chapter 5　The Rules of the Good Newtonian

1. A. M. Ampère, *Essai sur la philosophie des sciences,* vol. 1, Paris, 1834. For a partial translation of this work in Italian, see M. Bertolini, *Opere di André–Marie Ampère,* pp. 535–536, Turin, 1969.

2. A. M. Ampère, "Idées de Mr. Ampère sur la chaleur et sur la lumière," *Bibl. Univ. Genève* 49 (1832); idem, "Note de M. Ampère sur la chaleur et sur la lumière considérées comme résultant de mouvements vibratoires," *Ann. de Chimie et de Phys.* 58 (1835), *Bibl. Univ. Genève* 59 (1835), *Phil. Mag.* 7 (1835). These two papers are not included in the abovementioned Italian edition.

3. A. M. Ampère, "Note" (1835), 433.

4. Ibid., p. 437.

5. Ibid., p. 435.

6. Ibid., p. 440.

7. Ibid., pp. 441–442.

8. Ibid., p. 442.

9. Ibid., p. 443.

10. Ibid., p. 444.

11. S. D. Poisson, *Théorie mathématique de la chaleur,* Paris, 1835.

12. P. S. Laplace and A. L. Lavoisier, "Mémoire sur la Chaleur," *Mém. Acad. Sci.* (1780).

13. F. W. Bessel, "Versuche über die Kraft . . .," *Astr. Nachr.* 10 (1831), and *Pogg. Ann.* 25 (1832). For a French translation, see *Collections de mémoires relatifs à la physique publiés par la Société Française de Physique,* vol. 5, Paris, 1889.

14. A. Avogadro, "Nuove considerazioni sulle affinità de' corpi del calorico, calcolate per mezzo de' loro calori specifici, e de' loro poteri refrigenti allo stato gazoso," *Mem. Soc. Ital. Scienze* 19 (1823).

15. S. Carnot, *Réflexions sur la puissance motrice du feu et sur les machines propres à développer cette puissance,* Paris, 1824.

16. J. L. Gay-Lussac, *Leçons de physique à la Faculté des Sciences de Paris,* Paris, 1827.

17. J. B. Fourier, *Remarques générales sur l'application des principes de l'analyse algebrique aux équations trancendantes* (*Oeuvres,* vol. 2).

18. W. R. Hamilton, "On a General Method of Expressing the Paths of Light" (see note 67, chapter 3).

19. E. Clapeyron, "Mémoire sur la puissance motrice de la chaleur," *École Poly.* 14 (1834).

20. G. Lamé, *Cours de physique de l'École Polytechnique*, Brussels, 1836.

21. Quite interesting in this respect is C. Stewart Gillmor's *Coulomb and the Evolution of Physics and Engineering in Eighteenth-Century France*, Princeton, 1971.

22. Faraday was influenced by Wollaston's views. In a letter to de la Rive (12 September 1821) Faraday clearly expressed his skepticism about Ampère's hypotheses. See L. Pearce Williams, *The Selected Correspondence of Michael Faraday, Vol. I, 1812–1848*, Cambridge, 1971.

23. H. C. Oersted, "Experimenta circa effectum conflictus electrici in acum magneticum," *J. de Schweigger* 29 (1820), and *Ann. de Chimie et de Phisique* 34 (1820). See R. A. R. Tricker, *Early Electrodynamics. The First Law of Circulation*, Oxford, 1965.

24. H. C. Oersted, *Videnskaben om Naturens almindelige Love [The Science of the General Laws of Nature]*, Copenhagen, 1809; idem, *Recherches sur l'identité des forces chimiques et électrique*, Paris, 1813; idem, "Thermo-Electricity," in *Edinburgh Encyclopedia*. See: R. C. Stauffer, "Persistent Errors regarding Oersted's Discovery of Electromagnetism," *Isis* 44 (1953); idem, "Speculation and Experiment in the Background of Oersted's Discovery of Electromagnetism," *Isis* 48 (1957); L. Rosenfeld, "The Velocity of Light and the Evolution of Electrodynamics," *Nuovo Cimento Suppl. IV* 5 (1957); B. Dibner, *Oersted and the Discovery of Electromagnetism*, Waltham, Mass., 1962; F. Per Dahl, *L. Colding and the Conservation of Energy Principle*, New York, 1972.

25. A. M. Ampère, "Mémoire sur la théorie mathématique des phénomènes électrodynamiques uniquement déduite de l'expérience," *Mém. Acad. Sciences* 4 (1827). The memoir is dated 1825.

26. Quoted from the Introduction to *Opere di André-Marie Ampère*, p. 27.

27. It is interesting to note that in the 1826 edition published in Paris by Méquignon–Marvis there is no trace of the word *mathématique* in the title of the memoir, which is simply *Théorie des phenomenes* . . . .

28. *Opere di André-Marie Ampère*, p. 247.

29. Ibid., pp. 247–248.

30. Ibid., p. 250.

31. Ibid., p. 573.

32. Ibid., p. 354.

33. Ibid., p. 229 and foll.

34. Ibid., p. 339.

35. G. L. Lagrange, *Oeuvres*, vol. 14, p. 123, Paris, 1888.

36. C. A. Coulomb, "Recherches sur la meilleure manière de fabriquer les aiguilles aimantées," *Mém. Sav. Étrangers* 9 (1777) (see *Mémoires de Coulomb*, vol. 1 of *Collection de mémoires relatifs à la physique, publiés par la Société Française de Physique*, 8–9, Paris, 1884.

37. C. A. Coulomb, "Recherches théoriques et expérimentales sur la force de torsion et sur l'élasticité des fils de métal," *Mém. Acad. Royale des Sciences* (1784) (see *Mémoires de Coulomb*, pp. 65–103).

38. C. A. Coulomb, "Construction et usage d'une balance électrique fondée sur la propriété q'ont les fils de métal d'avoir une force de torsion proportionelle à l'angle de torsion," *Mémoires de Coulomb*, p. 110.

39. C. Stewart Gillmor, *Coulomb*, note 21, p. 176.

40. C. A. Coulomb, "Suite de recherches sur la distribution du fluide électrique entre plusieurs conducteurs. Détermination de la densité électrique dans les différents points de la surface de ces corps" (1788), *Mémoires de Coulomb*, pp. 230–272.

41. Ibid., p. 252.

42. C. A. Coulomb, "Du magnetism" (1789), *Mémoires de Coulomb*, pp. 273–318.

43. Ibid., p. 297: "Pour éviter toute discussion, j'avertis, comme je l'ai déjà fait dans les différents Mémoires qui précèdent, que toute hypothèse d'attraction et de répulsion suivant une loi quelconque non doit être regardée que comme une formule qui exprime un résultat d'expérience; si cette formule se déduit de l'action des molécules élémentaires d'un corps doué de certaines propriétés, si l'on peut tirer de cette première action élémentaire tous les autres phénomènes, si enfin les résultats du calcule théorique se trouvent exactement d'accord avec les mesures que fourniront les expériences, on ne pourra peut-être espérer d'aller plus loin, que lorsqu'on aura trouvé une lois plus générale qui enveloppe dans le même calcul des corps doués de différentes propriétés, qui, jusqu'ici, ne nous paraissent avoir entre elles aucune liaison."

44. Ibid., p. 302: "La conformité que nous trouvons ici entre les expériences fondamentales et le calcul semble donner un grand poids, soit à l'opinion de M. Oepinus, soit au système des deux fluides, telle que nous l'avons présentée; cependant il faut avouer qu'il y a quelques phénomènes qui semblent se refuser entièrement à ces hypothèses."

45. J. B. Biot, *Precis élémentaire de physique*, 3rd ed., Paris, 1823.

Chapter 6   The Empirical Ocean and the Conjectures of the Abbé Nollet

1. C. Truesdell, "A Program toward rediscovering the Rational Mechanics of the Age of Reason," *Arch. Hist. Exact Sci.* 1 (1960), 1.

2. S. D. Poisson, *Théorie mathématique de la chaleur*, Paris, 1835. According to Poisson, for instance, the mathematization of the theory of thermal phenomena takes place as follows: first a general hypothesis on the communication of heat, derived from experience and from analogy, is formulated; then all the consequences of the hypothesis are drawn by means of a "rigorous calculus"; "These consequences will thus be a transformation of the hypothesis itself, to which calculus neither adds nor subtracts anything"; finally, the consequences are compared with "the observed phenomena."

3. P. Duhem, *L'évolution de la mécanique*, Paris, 1905; idem, *La théorie physique, son objet et sa structure*, Paris, 1906. According to Duhem, the physics that Maxwell sets forth in the *Treatise on Electricity and Magnetism* is not a "logical system" but a series of "models." It is not true physics, just as "une galerie de tableaux n'est pas un enchaînement de syllogisme." Quoted from *La théorie physique*, p. 137.

4. P. Duhem, *La théorie physique*, p. 136.

5. J. A. Nollet, *Saggio intorno all'elettricitá de' corpi del Sig. abate Nollet*, p. 123, Venice, 1747 (Italian ed.).

6. Ibid., p. 5.

7. J. A. Nollet, *Recherches sur les causes particulières des phénomènes électriques*, Paris, 1749. The illustrations on pp. 148–149 have been taken from this volume, courtesy of the Biblioteca Comunale of Como.

8. Ibid., p. viii.

9. Ibid., pp. 73–74. It should be noted that when Nollet speaks of two matters he means two opposite fluxes: "L'électricité consiste . . . dans les deux mouvements contraires et simultanés de cette matière qu'on nomme électrique" (pp. 389–390).

10. Ibid., p. 57.

11. Ibid., p. 61.

12. Ibid., p. 65.

13. Ibid., pp. 63–64.

14. Ibid., p. 63.

15. Ibid., p. 66.

16. Ibid., p. 103.

17. Ibid., p. 106.

18. Ibid., p. 112.
19. Ibid., p. 267.
20. Ibid., p. 342 and foll.
21. Ibid., p. 355.
22. Ibid., p. 361.
23. Ibid., p. 363.
24. Ibid., pp. 365–366.
25. Ibid., pp. 381–382.
26. Ibid., pp. 389–390.
27. Ibid., p. 405 and foll.
28. Ibid., p. 32 and foll.
29. J. A. Nollet, *Leçons de physique expérimentale,* vol. 1, Paris, 1770.
30. Ibid., p. 184.
31. Ibid., pp. 180–181.
32. Ibid., pp. 159–162.
33. Ibid., "Préliminaire," pp. 1–5.
34. Ibid., pp. 159–160.
35. Ibid., p. 202.
36. B. Franklin, Letters II–III (1747), from *Experiments and Observations on Electricity, made at Philadelphia in America,* edited by I. B. Cohen, Cambridge, Mass., 1941.
37. Ibid., Letter V.
38. R. W. Home, "Franklin's Electrical Atmospheres," *British J. Hist. Sci.* 6 (1972).
39. F. Hauksbee, *Physico-Mechanical Experiments on various Subjects,* London, 1709–1719, edited by Douane H. D. Roller, New York, 1970.
40. H. Cavendish, "An Attempt to Explain Some of the Principal Phenomena of Electricity, by Means of an Elastic Fluid, *Phil. Trans.* 61 (1772).
41. Ibid., p. 584.
42. Ibid., p. 585.
43. Ibid.: "To explain this hypothesis more fully, suppose that 1 grain of fluid attracts a particle of matter, at a given distance, with as much force as n grains of any matter, lead for instance, repel it: then will 1 grain of electric fluid repel a particle of electric fluid with as much force as n grains of lead attract it; and 1 grain of electric fluid will repel 1 grain of electric fluid with as much force as n grains of lead repel n grains of lead."
44. Ibid., p. 586.
45. Ibid., pp. 587–588.

46. In a historical-critical analysis of Cavendish's work, Maxwell emphasizes the importance of Cavendish's law, according to which electric interaction varies as $1/r^2$ to a very good approximation. J. C. Maxwell, *The Electrical Researches of the Honourable Henry Cavendish*, Cambridge, 1879.

Chapter 7   The Paper World and Kelvin's Clouds

1. Galileo, *Dialogo sopra i due massimi sistemi del mondo*, Second Day, Salviati to Simplicius, VII, p. 139, 1632.

2. J. C. Maxwell, *Treatise on Electricity and Magnetism*, articles 552, 554, 570, and 574, Oxford, 1863.

3. L. Landau and E. Lifshitz, *Mécanique quantique. Théorie non relativiste*, p. 10, Moscow, 1966.

4. J. Tyndall, *Heat Considered as a Mode of Motion*, London, 1863.

5. The bibliography on such a complex problem is of course very extensive. See the original papers by Lorentz and Einstein, as well as the following works: G. Holton, "On the Origin of the Special Theory of Relativity," *Am. J. Phys.* 28 (1960); S. Goldberg, "H. Poincaré and Einstein's Theory of Relativity," *Am. J. Phys.* 35 (1967); G. Holton, "Einstein, Michaelson and the 'Crucial Experiment'," *Isis* 60 (1969); R. McCorrmach, "H. A. Lorentz and the Electromagnetic View of Nature," *Isis* 61 (1970); A. I. Miller, "A Study of H. Poincaré's 'Sur la Dynamique de l'Electron'," *Arch. Hist. Exact Sci.* 10 (1973); E. Zahar, "Why did Einstein's Programme Supersede Lorentz's?," *Brit. J. Phil. Sci.* 24 (1973).

6. M. Faraday, "On the Conservation of Force," *Proc. Royal Soc.* 2 (1857).

7. Ibid.

8. P. S. Laplace, *Traité de mécanique céleste*, vol. 5, chapter 12, Paris, 1825.

9. F. Delaroche, J. E. Bérard, "Sur la détermination de la chaleur spécifique de differents gaz," *Ann. de Chimie*, (1813), 72–110, 113–182.

10. J. B. Fourier, *Théorie analytique de la chaleur* (1822) (see J. B. Fourier, *Oeuvres*, edited by G. Darboux, Paris, 1890.

11. Ibid., p. 110 and foll.

12. See works by A. M. Ampère cited in note 2, chapter 5.

13. S. G. Brush, "The Wave Theory of Heat: A Forgotten Stage in the Transition from the Caloric Theory to Thermodynamics," *British J. Sci.* 5 (1970), 18.

14. J. P. Joule, "On the Mechanical Equivalent of Heat," *Phil. Trans.* 61 (1850). Interesting considerations on the question of academic mentality in relationship to J. J. Waterston's monograph, rejected in 1845 as

"nonsense," or to J. Hearapath's works can be found in C. Truesdell's *Essays in the History of Mechanics,* Berlin, 1968.

15. W. Thomson (Lord Kelvin), "An Account of Carnot's Theory of the Motive Power of Heat," *Trans. Royal Soc. Edinburgh* (1849).

16. W. J. M. Rankine, letters, *Ann. Physik und Chemie* 9 (1850), *Phil. Mag.* 2 (1851). In addition: "On the Hypothesis of Molecular Vortices, or Centrifugal Theory of Elasticity, and Its Connection with the Theory of Heat," *Phil. Mag.* 10 (1855), 67. Also interesting is a letter to Joule, *Phil. Mag.* 5 (1853), 29. Rankine's knowledge of the theories of Laplace and Poisson is attested by two of his papers: "On Laplace's Theory of Sound" and "On Poisson's Investigation of the Theory of Sound," *Phil. Mag.* 1 (1851), 5.

17. W. J. M. Rankine, "On the Mechanical Action of Heat, Especially in Gases and Vapours," *Phil. Mag.* 7 (1854), 42.

18. W. Thomson (Lord Kelvin), "An Account of Carnot's Theory of the Motive Power of Heat," *Trans. Royal Soc. Edinburgh* (1849).

19. The first step toward the solution of this theoretical knot comes in 1850 with Clausius' mathematical-physical studies. Certain fundamental differences between Clausius' and Kelvin's projects will not be resolved, however. On the contrary, they become more serious with the years and provoke heated debates.

20. "From the everyday intuitions and debates about determinism, mechanism and the like there has emerged slowly a tradition of conceptual analysis in which the ideal is to characterize conceptually interesting questions so sharply that they admit of formally precise answers. (Thus 'Is the theory deterministic?' becomes 'Does the theory admit such and such mathematical construction?' . . .)." Quoted from the Preface to *The Logico-Algebraic Approach to Quantum Mechanics,* vol. 1, edited by C. A. Hooker, Dordrecht, 1975.

21. W. Ostwald, *Die Überwindung des wissenschaftlichen Materialismus,* Lübeck, 1895. Translated in 1895 in the twenty-first issue of the *Revue Générale des Sciences,* Ostwald's work elicited some interesting answers from M. Brillouin and A. Cornu.

22. W. Thomson (Lord Kelvin), "Kinetic Theory of the Dissipation of Energy," *Nature* (9 April 1874). Kelvin writes: "In abstract dynamics the instantaneous reversal of the motion of every moving particle of a system causes the system to move backwards, each particle of it along its old path, and at the same speed as before, when again in the same position. That is to say, in mathematical language, any solution remains a solution when t

is changed into −*t*. In physical dynamics this simple and perfect reversibility fails." See also W. Thomson (Lord Kelvin), "The Sorting Demon of Maxwell," *Proc. Royal Inst.* 9 (1879) (*Popular Lectures and Addresses,* vol. 1, pp. 137–141, London, 1889), and the series of articles by J. Loschmidt cited in note 25, chapter 4. For further developments of the problem of irreversibility see: E. Zermelo, "Über einen Satz der Dynamik und der mechanishen Wärmetheorie," *Wied. Ann.* 57 (1896); L. Boltzmann, "Entgegnung auf die wärmetheoretischen Betrachtungen irreversibler Vorgänge. Eine Antwort auf Hrn. Boltzmanns 'Entgegnung'," *Wied. Ann.* 59 (1896); idem, "Zu Hrn. Zermelo Abhandlung 'Ueber die mechanischen Erklärungen irreversibler Vorgänge," *Wied. Ann.* 60 (1897); idem, "Ueber einen mechanischen Satz Poincarés," *Wied. Ann.* 61 (1897); idem, *Vorlesungen über Gastheorie,* vols. 1, 2, Leipzig, 1895–1898. For an English translation of Boltzmann's major works on the theory of gases, see L. Boltzmann, *Lectures on Gas Theory,* translated and edited by S. Brush, Berkeley, 1964.

23. W. Thomson (Lord Kelvin), "Nineteenth Century Clouds over the Dynamical Theory of Heat and Light," allocution to the Royal Institution, 27 April 1900.

24. L. Boltzmann, "Studien über das Gleichgewicht der lebendigen Kräfte zwischen bewegten materiellen Punkten," *Sitz. König. Akad. Wien* 58 (1868); J. C. Maxwell, "On Boltzmann's Theorem on the Average Distribution of Energy in a System of Material Points," *Cambridge Phil. Soc. Trans.* 3 (1879); W. Strutt (Lord Rayleigh), "The Law of Partition of Kinetic Energy," *Phil. Mag.* 49 (1900).

25. W. Strutt (Lord Rayleigh), "Remarks upon the Law of Complete Radiation," *Phil. Mag.* 49 (1900).

26. W. Wien, "Über die Energievertheilung in Emissionsspectrum eines schwarzen Körpers," *Wied. Ann.* 58 (1896), and *Phil. Mag.* 43 (1897).

27. W. Strutt (Lord Rayleigh), 1902 notation on p. 485 of vol. 4 of *Scientific Papers, 1892–1901,* Cambridge, 1903.

28. M. Polanyi, "Science and Reality," *British J. Phil. Sci.* 18 (1967), 195: "To see a good problem is to see something hidden and yet accessible. This is done by integrating some raw experiences into clues pointing to a possible gap in our knowledge. To undertake a problem is to commit oneself to the belief that you can fill in this gap and make thereby a new contact with reality. Such a commitment must be passionate; a problem which does not worry us, that does not excite us, is not a problem; it does not exist."

A Chronology of Physics in the Second Scientific Revolution

In order to put the subjects discussed in this book in the proper historical perspective, we have prepared a schematic chronology of the most significant events that marked the transition from classical mechanics to the modern view of the physical world up to the threshold of the twentieth century.

1687 Newton's *Philosophiae naturalis principia mathematica* are published.

1690 C. Huygens formulates the wave theory of light in his *Traité de la Lumière*. Leibniz writes the *Fundamenta calculi logici*.

1704 Newton's *Optics* is published.

1724 The Newtonian W. J. s'Gravesande discusses the relationship between physics and mathematics in *De evidentia*.

1737 Scientific expeditions to Lapland and Peru confirm Newton's theoretical predictions about the shape of the earth.

1743 J. B. le Rond d'Alembert's *Traité de dynamique* is published.

1759 Three fundamental works appear in print: *Philosophiae naturalis theoria* by R. G. Boscovich; *Tentamen theoriae electricitatis et magnetismi* by U. T. Aepinus; *Theoria motus corporum* by L. Euler.

1777 C. A. Coulomb establishes new bases for the study of electricity and magnetism.

1788 First edition of G. L. Lagrange's *Mécanique analytique*.

1789 A. L. Lavoisier publishes his *Traité élémentaire de chimie*.

1795 J. Hutton's *Theory of the Earth* is printed.

1799 P. S. Laplace completes the first two volumes of the *Traité de mécanique céleste*.

1800 A. Volta's paper on electricity by contact.

1807 H. Davy's lecture "On Some Chemical Agents of Electricity."

1809 Publication of *Philosophie zoologique* by J. B. Lamarck.

1811 A. Avogadro's molecular hypothesis.

1814   Laplace confronts the problem of physical mathematical knowledge in the *Essai philosophique sur les probabilités*. K. F. Gauss publishes his *Methodus nova integralium valores per approximationem inveniendi*.

1815   G. Cuvier's *Discours sur les révolutions de la surface du globe* is published.

1818   J. A. Fresnel's essay on ether and optical phenomena.

1819   P. L. Dulong's and A. T. Petit's experiments on the specific heat of solids.

1820   J. C. Oersted's experiments on the interactions between electrical currents and magnetic needles. A. M. Ampère's first works on electrodynamics.

1822   Publication of the *Théorie analytique de la chaleur* by J. B. Fourier.

1823   N. I. Lobachevski's and J. Bolyai's studies on non-Euclidean geometries.

1824   S. Carnot publishes the *Réflexions sur la puissance motrice du feu*.

1825   Fifth and last volume of the *Traité de mécanique céleste*.

1827   Death of Laplace.

1830   A. Cauchy's molecular theory of ether.

1831   M. Faraday begins his *Experimental Researches in Electricity*.

1834   W. R. Hamilton writes the *First Essay on a General Method in Dynamics*.

1838   After explaining electromagnetic induction, formulating hypotheses on the lines of force and reshaping electrochemistry, Faraday criticizes in depth the concept of action at a distance.

1842   K. G. Jacobi generalizes Hamilton's method. J. P. Joule and R. J. Mayer start research in the conservation of energy. Discovery of the Doppler effect.

1845   Faraday's experiments in magneto-optics.

1846   W. Weber extends the study of electrodynamics.

1847   First edition of H. Helmholtz's *Über die Erhaltung der Kraft*.

1849   A. H. L. Fizeau's experiment on the speed of light.

1850   Modern thermodynamics is born: Joule publishes *On the Mechanical Equivalent of Heat* and R. Clausius *Über die bewegende Kraft der Wärme*.

1851   L. Foucault's experiment on the earth's rotational motion.

1852   W. Thomson (Lord Kelvin) enunciates the principle of degradation of mechanical energy. Faraday accuses Newtonian physics of dogmatism.

1854   Publication of G. Boole's *An Investigation of the Laws of Thought*.

1855   W. M. Rankine writes *Outlines of the Science of Energetics*.

1856   Probabilistic models of molecular motion by A. Krönig and Clausius.

1859   G. R. Kirchhoff builds the spectroscope. C. Darwin's *On the Origin of Species* is published in London. J. Plücker studies cathode rays.

1860   J. C. Maxwell publishes *Illustrations on the Dynamical Theory of Heat*.

1861   Kirchhoff formulates the concept of blackbody. Maxwell analyzes the concept of electromagnetic field. H. A. Lorentz studies the concept of delayed potential.

1865   Clausius elaborates the details of the concept of entropy. Maxwell publishes the monograph *A Dynamical Theory of the Electromagnetic Field*.

1866   Riemann dies. The kinetic theory of gases is enunciated by Maxwell in the essay *On the Dynamical Theory of Gases*.

1867   Faraday dies. Riemann's monograph on the foundations of geometry is published. Kelvin writes *On Vortex-Atoms*.

1868   L. Boltzmann studies equipartition of molecular energy. The concept of mass is criticized by E. Mach. H. Helmholtz analyzes the bases of geometry.

1872   Boltzmann develops the first demonstration of the H theorem and proposes the quantum of molecular energy. F. Klein presents the celebrated "Erlangen program" on the relationship between geometry and group theory.

1873   J. D. Van der Waals' equations for real gases. First edition of Maxwell's *Treatise on Electricity and Magnetism*.

1875   New experiments by W. Crookes on cathode rays.

1878   G. Cantor's first writings on infinite sets.

1879   A. Einstein is born. Maxwell dies. Stefan's law is enunciated.

1880   The controversy between the wave and corpuscular theories of cathode rays sharpens. Crookes' corpuscular model seems to be invalidated by the experiments of G. H. Wiedemann and E. Goldstein.

1881   A. A. Michelson performs the first interferometric experiment.

1883   First edition of E. Mach's book on the historical-critical development of mechanics. Hertz's wave theory of cathode rays.

1884   Boltzmann demonstrates the validity of Stefan's law.

1887   Michelson–Morley experiment on the earth's motion. Hertz's experiments on electromagnetic waves. M. Planck criticizes molecular models.

1889   H. Poincaré publishes his essay *Sur le problème des trois corps et les équations de la dynamique*. The wave model of radiation by W. Strutt (Lord Rayleigh).

1890   Hertz's memoirs on electrodynamics.

1892   H. A. Lorentz's theory of the electron.

1894   Publication of Hertz's *Die Prinzipien der Mechanik*. Hertz dies. A general debate on irreversibility begins in the magazine *Nature*.

1895   At the Congress of Lübeck Boltzmann's theses are defeated by the philosophy of Ostwald, Mach, and Helm.

1896   Discovery of X rays and Becquerel radiation. Theoretical analysis of radiation leads to Wien's law. The Zeeman effect.

1897   J. J. Thomson determines the charge-to-mass ratio and the speed of electrons. Planck re-elaborates Wien's law and sharply criticizes energetics.

1899   Lorentz refines his studies on the Lorentz transformation and formulates the idea of local time. Thomson delivers a lecture "On the Existence of Masses Smaller than Atoms." E. Rutherford carries out research on α and β rays. M. and P. Curie investigate polonium and radium. H. Haga and C. H. Wind study the phenomenon of electron scattering. Thomson and Lenard work on photoelectric radiation. Planck's research program moves closer to Boltzmann's theses.

1900   Rayleigh enunciates a new formula for radiation. Planck introduces the H constant. P. Villard investigates γ rays. Poincaré re-examines Lorentz's theory and publishes *La théorie de Lorentz et le principe de réaction*, raising questions of compatibility between the theory of electrons and mechanics. J. Larmor's *Ether and Matter* is published. Lord Kelvin delivers an allocution on the clouds in physics. The problem of atomic structure becomes all-important.

1904   At the Congress of Arts and Sciences in St. Louis H. Poincaré presents the first enunciation of the principle of relativity.

1905   Three essays by Einstein, dated March, May, and June, respectively, appear in the *Annalen der Physik*. The first article discusses the concept of quanta of light and explains the photoelectric effect. The second contains a theory of Brownian motion and arguments in support of the existence of atoms. The third introduces the theory of relativity starting from asymmetries in Lorentz's theory.

1906   Boltzmann commits suicide. The philosophical battle over the physical sciences sharpens with the publication of P. Duhem's *La théorie physique*, followed a few months later by H. Bergson's *L'évolution créatrice*.

1908   Poincaré publishes *Science et méthode*. Thomson perfects his model of an atom without nucleus. Rutherford sets the foundations for the experiments on the scattering of α particles by thin foils. II. Minkovski defends Einstein's theory.

1911   Rutherford proposes an atom with nucleus. The Solvay conference on radiation and quanta takes place in Brussels. Thesis dissertation by N. Bohr.

1913   Bohr's article "On the Constitution of Atoms and Molecules" appears in *Philosophical Magazine*.

Index